The Economics of Energy Policy in China

NEW HORIZONS IN ENVIRONMENTAL ECONOMICS

General Editor: Wallace E. Oates, *Professor of Economics, University of Maryland*

This important series is designed to make a significant contribution to the development of the principles and practices of environmental economics. It includes both theoretical and empirical work. International in scope, it addresses issues of current and future concern in both East and West and in developed and developing countries.

The main purpose of the series is to create a forum for the publication of high quality work and to show how economic analysis can make a contribution to understanding and resolving the environmental problems confronting the world in the late twentieth century.

Recent titles in the series include:

The Economic Theory of Environmental Policy in a Federal System
Edited by John B. Braden and Stef Proost

Environmental Taxes and Economic Welfare
Reducing Carbon Dioxide Emissions
Antonia Cornwell and John Creedy

Economics of Ecological Resources
Selected Essays
Charles Perrings

Economics for Environmental Policy in Transition Economies
An Analysis of the Hungarian Experience
Edited by Péter Kaderják and John Powell

Controlling Pollution in Transition Economies
Theories and Methods
Edited by Randall Bluffstone and Bruce A. Larson

Environments and Technology in the Former USSR
Malcolm R. Hill

Pollution and the Firm
Robert E. Kohn

Climate Change, Transport and Environmental Policy
Empirical Applications in a Federal System
Edited by Stef Proost and John B. Braden

The Economics of Energy Policy in China
Implications for Global Climate Change
ZhongXiang Zhang

Advanced Principles in Environmental Policy
Anastasios Xepapadeas

Taxing Automobile Emissions for Pollution Control
Maureen Sevigny

Global Environmental Change and Agriculture
Assessing the Impacts
Edited by George Frisvold and Betsey Kuhn

Fiscal Policy and Environmental Welfare
Modelling Interjurisdictional Competition
Thorsten Bayindir-Upmann

The Economics of Energy Policy in China

Implications for Global Climate Change

ZhongXiang Zhang
Research Fellow, Department of Economics and Public Finance, University of Groningen, The Netherlands and Professor of Economics, Centre for Environment and Development, Chinese Academy of Social Sciences, Beijing, China

NEW HORIZONS IN ENVIRONMENTAL ECONOMICS

Edward Elgar
Cheltenham, UK • Northampton, MA, USA

Published by
Edward Elgar Publishing Limited
8 Lansdown Place
Cheltenham
Glos GL50 2HU
UK

Edward Elgar Publishing, Inc.
6 Market Street
Northampton
Massachusetts 01060
USA

A catalogue record for this book
is available from the British Library

Library of Congress Cataloguing in Publication Data

Zhang, ZhongXiang, 1963–
     The economics of energy policy in china: implications for global
 climate change / ZhongXiang Zhang.
          (New horizons in environmental economics)
     Includes bibliographical references and index
         1. Energy policy—Environmental aspects—China. 2. Atmospheric
 carbon dioxide—Economic aspects—China—Mathematical models.
 3. Environmental economics—Mathematical models. 4. Greenhouse
 effect. Atmospheric—Economic aspects. I. Title. II. Series.
 HD9502.C62Z46 1997
 333.79'0951—dc21   97–25030
                                                          CIP

ISBN 1 85898 614 1

Printed and bound in Great Britain by
MPG Books Ltd, Bodmin, Cornwall

Contents

Figures

Tables

Preface

My interest in and the choice of working on an economic analysis of energy and environmental policy are inspired by at least three factors. First, what impressed me when I arrived in Holland in 1989 was that, although I knew that the Netherlands had a good reputation in the environmental area, the Dutch political parties viewed climate change as one of the 1989 election issues. Second, Dutch economists have long been internationally renowned for their work. Third, systematic and comprehensive research on the economy-energy-environment system and its interaction is still lacking in China. The first two factors mean that the Netherlands is an ideal place for doing research on environmental economics, while the third factor highlights the necessity of doing such research for China. This has motivated me to propose the project entitled 'Compatibility of CO_2 Emission Reduction Targets with Long-term Economic Development in China'. The project has been funded by the Netherlands National Research Programme on Global Air Pollution and Climate Change (NOP) under contract NOP-852064, of which my PhD dissertation was the result. This book has grown out of the dissertation. Turning it into the book has mainly consisted of some rearrangement of the material, the rewriting of some passages, the omission of some others, as well as some updating of the material.

For the completion of the book, I owe a great deal to various people. I am most indebted to Prof. Henk Folmer (University of Wageningen). His willingness to be my promotor pleasantly surprised me since he served as the President of the European Association of Environmental and Resource Economists at the time I contacted him. His comments and criticisms have contributed to the successful completion of my dissertation. Moreover, he has given me the freedom necessary to develop my own style and perspective on the research. My sincere thanks are also due to my other promotor Prof. Paul van Beek (University of Wageningen). He has helped me to get my PhD research started in Wageningen. His detailed comments have made the presentation of mathematical formulas in Chapter 8 both clearer and more precise.

Prof. Ruud Lubbers (Tilburg University), the former Dutch Prime Minister, has shown great interest in the research, and I feel honoured to have received his comments. I also benefitted from Prof. Ekko van Ierland (University of Wageningen). He has participated in some of the regular discussion meetings and carefully reviewed all the final manuscripts of my dissertation. Furthermore, I would like to express my thanks to Dr Jos Bruggink for his contribution to the draft project proposal at the time I was working at the Policy Studies Department of the Netherlands Energy Research Foundation (ECN). My thanks also go to Tom Kram at ECN for keeping me informed of the IEA-ETSAP activities.

During the course of the research I have been fortunate to receive support, in one way or another, from my Chinese colleagues from the State Planning Commission, Ministry of Electric Power, Chinese Academy of Social Sciences, Chinese Academy of Sciences, National Environmental Protection Agency, and China Energy Research Society. I especially wish to thank Prof. Hu Zhaoyi, Prof. Yao Yufang, Prof. Zhang Zhengmin, Prof. Tang Yuan, Prof. Li Wei-zheng, Prof. Deng Shuhui, Prof. Ye Ruqiu and Prof. Wang Qingyi for arranging my field work in Beijing and for providing me with many scarce and otherwise unavailable data as well as with useful suggestions.

I benefitted a great deal from discussions with Prof. Lars Bergman at the Stockholm School of Economics. He has helped me to interpret the calculation results. I would like to express my thanks to Prof. Alan Manne at the Stanford University for providing me with necessary information about GLOBAL 2100 for the purpose of model comparison. I am also grateful to Prof. Leen Hordijk (University of Wageningen), Prof. Catrinus Jepma (Universities of Amsterdam and Groningen), Prof. Ekko van Ierland and Prof. Kees van Kooten (University of British Columbia at Vancouver) for their willingness to participate in the PhD examination committee. Furthermore, my special thanks are due to the Department of General Economics for its hospitality and for providing an ideal research environment. My gratitude also goes to Mrs. Diny Dijkhuizen for correcting the English text of the dissertation.

Financial support from the NOP is gratefully acknowledged. I would also like to express my thanks for a fellowship granted by the University of Wageningen and the funding of another project provided by the Netherlands Ministry of Housing, Spatial Planning and the Environment under contract 95140042. The former has offered me the opportunity to propose the NOP project, while the latter has made it possible to finalize the book.

Needless to say, I remain responsible for all the views expressed in this book and any errors and omissions that may remain.

Finally, as far as my family is concerned, I would like to thank my parents, brothers and sisters for encouraging me to take up the PhD research. I regret very much that my father has not been able to see the completion of this book. I dedicate it to him. Furthermore, I would like to express my special thanks to my wife Li Guoxin, who quitted office at the Chinese Academy of Social Sciences in Beijing, for her enduring patience and affection during the long journey to completing my dissertation on economics. She made home hard to leave and good to return to.

ZhongXiang Zhang
Wageningen, The Netherlands

1 Introduction

1.1 Background and aims

In recent years, there has been growing concern about global warming resulting from increased atmospheric concentrations of the so-called greenhouse gases and the resulting socioeconomic impacts. Although there are still uncertainties regarding the magnitude, timing and regional patterns of climate change, there is a growing consensus in scientific and policy-making circles that climate change and instability, including a rise in global atmospheric temperatures, a change in frequency and severity of storms, shifts in precipitation patterns, and a rise in sea level, are most likely over the next century.

The greenhouse gases (GHG) are carbon dioxide (CO_2), chlorofluorocarbons (CFCs), methane (CH_4), tropospheric ozone (O_3) and nitrous oxide (N_2O). These GHG emissions derive from a number of human activities, including energy production and use, non-energy industrial processes (primarily the production and use of CFCs), deforestation and agricultural practices. Estimations of the relative contributions to global warming between 1980 and 2030 arising from these activities and GHG are shown in Table 1.1, assuming that current trends are to continue.

It can be seen that CO_2 emissions alone contribute about half of the global warming over the 40 years, and thus are the major cause of the greenhouse effect. CFCs are the other principal contributors. With respect to the protection of the stratospheric ozone layer, however, CFCs are subject to control and thus are expected to be phased out, although their effect will persist for a considerable time. This implies that the contribution of CO_2 to global warming is likely to rise over the next decades with the phase-out of the CFCs (Nordhaus, 1991a). The CH_4 and N_2O emissions arise from a range of sources, and the relative importance of these sources remain still uncertain. Moreover, it is generally thought that CH_4 and N_2O, particularly those emissions from agriculture, are difficult to measure.

When the contributions by the various sources of GHG are considered, energy production and use, with which all GHG with the exception of the CFCs are associated to a greater or lesser extent, are expected to contribute about half of the increased greenhouse effect, while deforestation and agriculture together contribute 25%, with industry being responsible for the remaining 25% (World Resources Institute, 1990).

Given that carbon dioxide is the greatest contributor to global warming and that the burning of fossil fuel is an important source of GHG emissions, a number of proposals have been put forward for the limitation of global GHG emissions, with CO_2, particularly the fossil-fuel related emissions, being set the main target.

Recognizing the great difficulties in reducing CO_2 emissions, for instance, the Toronto Conference on the Changing Atmosphere has called for only a 20% cut by 2005 and a 50% cut by 2025 in global CO_2 emissions relative to their 1988 levels, although a 50-80% cut in global CO_2 emissions is thought to be required to stabilize the atmospheric CO_2 concentration (UNEP, 1988).

Table 1.1   *Estimated contributions to global warming by greenhouse gases and human activities between 1980 and 2030*

	CO_2	CFCs	CH_4	O_3	N_2O	% warming by sector
Energy	35	-	4	6	4	49
Deforestation	10	-	4	-	-	14
Agriculture	3	-	8	-	2	13
Industry	2	20	-	2	-	24
% warming by gas	50	20	16	8	6	100

Source: World Resources Institute (1990).

At present the industrialized countries are responsible for the majority of global CO_2 emissions, and must bear the major burden of the carbon reduction. This is why the United Nations Framework Convention on Climate Change at the Earth Summit in Rio de Janeiro commits industrialized countries parties to cut down emissions of CO_2 and other greenhouse gases to their 1990 levels by the year 2000 (Grubb and Koch *et al.*, 1993).[1] This goal has become enshrined as the acid test of whether the industrialized world is serious about tackling global warming. Efforts by the industrialized countries, however, may well be over-shadowed by increases in CO_2 emissions from the developing countries because these countries are the fastest growing emissions sources. Looking at a shift in the pattern of regional contributions, North America and Western Europe, together accounting for 68% of total world emissions in 1950, represented only about 38% of the expanded total in 1987. By contrast, the portion attributable to China and other developing countries in Latin America, Africa and Southern Asia grew from just 7% to 28% of the world total during the corresponding period (Kats, 1990). This trend highlights the fact that the developing countries must be part of the solution. The dire prediction of future global CO_2 emission increases cannot otherwise be avoided if the developing countries industrialize along the same path as the developed counterparts have done.

---

[1] The actual wording in the Convention is convoluted and deliberately ambiguous in key parts, owing to the refusal of the Bush administration to accept clear legal obligations to emission targets in the treaty. See Grubb and Koch *et al.* (1993) for a short overview of that Convention.

Being a developing country, China is currently undergoing a significant transformation. This has led to spectacular growth of the Chinese economy, with an annual growth of about 9% for GNP during the period 1980-90. In the meantime, energy consumption rose from 602.75 Mtce (million tons of coal equivalent) in 1980 to 987.03 Mtce in 1990. The corresponding CO_2 emissions grew from 358.60 MtC (million tons of carbon) to 586.87 MtC during the same period (Zhang, 1994b). This means that China ranks second in global CO_2 emissions behind the US if the Soviet emissions are distributed over the new independent republics. Under a business-as-usual scenario, China's contribution to global CO_2 emissions is estimated to rise from 11% in 1990 to 17% in 2050 and to 28% in 2100 (Manne, 1992). Given that China's economic development depends mainly on its domestic market, its rapid growth may remain decoupled from the medium or slow growth for the rest of the world. This possibility, combined with industrialized countries succeeding in stabilizing their CO_2 emissions, would make China's share in global CO_2 emissions much larger than the above mentioned. Thus, advocates of controlling CO_2 emissions call for substantial efforts in China. Indeed, given that China is the world's most populous country and largest coal producer and consumer, its coal-dominated energy structure and energy-and-carbon-intensive economy, its economic development and efforts to limit CO_2 emissions are of great influence on future global CO_2 emissions.

Because of the global characteristics of climate change and China's potential importance as a source of CO_2 emissions, there are global models, though relatively few, that cover various political-economic regions and treat China as a separate region. These models are of a wider regional scope and are thus able to give insight into the regional effects of a variety of international CO_2 agreements on international trade and welfare consequences, including impacts on China. However, they suffer from a lack of sectoral information, partly because of unavailability of data and computational problems. Combined with the fact that the effectiveness of international agreements would depend eventually on decisions that reflect national priorities, this suggests that for countries with a large proportion of CO_2 emissions like China, a country-specific model should be developed. The single-country (China) model should allow for a more detailed and reliable analysis than the existing global models in terms of sectoral scope and energy sources. We believe that such an analysis at national level is useful to broaden the picture painted by global models. It is also useful to serve as a check or even as a complement of the results obtained through global models.

Against this background, the project 'Compatibility of CO_2 Emission Reduction Targets with Long-term Economic Development in China' was initiated. The purposes of this study were threefold:

1. to provide an analysis of the Chinese energy system in order to shed light on its implications for China's future CO_2 emissions;
2. to provide a macroeconomic analysis of CO_2 emission limits for China, using a newly-developed computable general equilibrium model of the Chinese economy; and

3.   to provide a cost-effective analysis of carbon abatement options in China's electricity sector by means of a technology-oriented dynamic optimization model.

Through these analyses, the following important results could be obtained: the effects of CO_2 emission limits for China at both sectoral and macroeconomic levels, the effects of recycling carbon tax revenues, the magnitude of carbon taxes across regions in order to achieve the same percentage of carbon reductions relative to the baseline, the levelized cost of generation and the marginal cost of CO_2 reduction among power plants, and the impacts of compliance with CO_2 limits in China's power industry. These results can contribute to forming national, cost-effective response strategies for climate change and a necessary basis for China's development of joint implementation projects with other countries.

1.2   Outline of the book

This study is the first systematic and comprehensive attempt to deal with the economic implications of carbon abatement for the Chinese economy in the light of the economics of climate change, of which this book is the result.

The book consists of nine chapters. Prior to analysing in detail the Chinese energy system and quantifying the economy-wide effects of limiting China's CO_2 emissions, Chapter 2 first discusses some economic aspects of climate change, including the consequences of climate change, some damage estimates for a doubling of atmospheric CO_2 concentration, strategies for responding to climate change, and policy instruments for the control of CO_2 emissions. This in turn will serve as a good guide to pursuing the case study for CO_2 emissions in China.

Chapter 3 deals with the analysis of the Chinese energy system in the CO_2 context. The issues covered include China's energy resources and their development, the Chinese energy consumption patterns, the achievements and remaining problems in China's electricity sector, China's energy conservation in an international perspective, the analysis of historical CO_2 emissions in China, and environmental challenges for the Chinese energy system. Examining these aspects of the Chinese energy system is not only to contribute to a better understanding of the Chinese energy system, but also to shed light on its implications for China's future CO_2 emissions.

Chapter 4 is concerned with the alternative economic modelling approaches to cost estimates for limiting CO_2 emissions. The approaches discussed include the *ad hoc* approach, dynamic optimization approach, input-output approach, macroeconomic approach, computable general equilibrium (CGE) approach, and the hybrid approach. Without going into too much detail, this chapter gives an assessment of the relative strengths and weaknesses of these different economic modelling approaches. Its purpose is to illustrate how these different approaches are able to shed light on different aspects of cost estimates for the control of CO_2

emissions and, at the same time, to show the theoretical rationale for choosing a CGE approach for a macroeconomic analysis of CO_2 emission limits in China and for linking a CGE model of the Chinese economy with a power planning model of China's electricity sector.

For macroeconomic analysis of CO_2 emission limits for China, a time-recursive dynamic CGE model of the Chinese economy has been designed and is described in Chapter 5. This CGE model operates by simulating the operation of markets for factors, products and foreign exchange, with equations specifying supply and demand behaviour across all markets. The model includes ten producing sectors and distinguishes four energy inputs. The model is made up of the following blocks: production and factors, prices, income, expenditures, investment and capital accumulation, foreign trade, energy and environment, welfare, and market clearing conditions and macroeconomic balances. Moreover, the CGE model, which is confined to the economic impacts of carbon emission limits, highlights the relationships between economic activity, energy consumption and CO_2 emissions. Thus, the model makes it possible to analyse the Chinese economy-energy-environment system interactions simultaneously, at both sectoral and macroeconomic levels in general, and the economic impacts of limiting CO_2 emissions in particular for this study. The model is also able to calculate the welfare impacts of carbon abatement policies. Furthermore, the CGE model incorporates an explicit tax system. This makes it suitable for estimating the 'double dividend' from the imposition of a carbon tax that is incorporated as a cost-effective means of limiting CO_2 emissions.

Chapter 6 deals with the data requirements of the CGE model, the calibration of the model's parameters, and solution approaches. These issues are considered to be essential for empirical application of the CGE model.

After the detailed description of the CGE model in Chapter 5 and some essential work done for empirical application of the model in Chapter 6, Chapter 7 is devoted to analysing the economy-wide impacts of alternative carbon limits through counterfactual policy simulations. For this purpose, a business-as-usual scenario is first developed. Counterfactual policy simulations are then carried out to compute the macroeconomic and sectoral implications of two alternative carbon limits relative to the business-as-usual scenario and to determine the efficiency improvement of four indirect tax offset scenarios relative to the tax retention scenarios. Moreover, a comparison with other studies for China, including the well-known global studies based on GLOBAL 2100 and GREEN (Manne, 1992; Martins *et al.*, 1993), is made in terms of both the baseline scenarios and the carbon constraint scenarios.

Chapter 8 attempts to shed light on technological aspects of carbon abatement in China's power industry and is thus devoted to satisfying energy planning requirements. To that end, the chapter first describes the MARKAL model (Fishbone *et al.*, 1983), from which our technology-oriented dynamic optimization model for power system expansion planning is adapted. This is followed by a description of 15 types of power plants in terms of their technical, economic and environmental characteristics, and by a comparison of these plants in terms of both the levelized cost of generation and the marginal cost of CO_2 reduction. Driven by the baseline electricity demands that are estimated by the CGE model, the model is then used to develop the business-as-usual scenario for China's electricity supply and to analyse the impacts of compliance with CO_2 limits in the power industry.

Finally, Chapter 9 summarizes the conclusions of this study and points out some areas where there is a need for further methodological and empirical work to enrich the policy relevance of this study.

2 Some Economic Aspects of Climate Change[1]

2.1 Introduction

Prior to analysing the Chinese energy system in detail and quantifying the economy-wide effects of limiting China's CO_2 emissions, I will first discuss some economic aspects of climate change in general. This in turn will serve as a good guide to pursuing the case study for CO_2 emissions in China. Section 2.2 describes the consequences of climate change. Section 2.3 discusses some damage estimates for a doubling of atmospheric CO_2 concentration. In Section 2.4 strategies to respond to climate change are briefly touched upon, while alternative economic modelling approaches to cost estimates for limiting CO_2 emissions are assessed in Chapter 4. Section 2.5 gives an overview of policy instruments targeted at the control of CO_2 emissions, including the command-and-control approach, energy taxes, carbon taxes, tradeable carbon permits and joint implementation. In the overview, special attention is paid to the economic instruments. The chapter ends with some concluding remarks.

2.2 Consequences of climate change

Natural ecosystems are already stressed by the growth of population, industrial development, the need for agricultural land, and the unsustainable exploitation of natural resources. The resulting consequences include air and water pollution, deforestation, ground water withdrawal and soil erosion. On top of these environmental stresses, global warming and associated climate change will bring the additional consequences. The combination of climate change with these stresses could be more dangerous than any one of them taken by itself (Ayres and Walter, 1991).

The great bulk of the scientific work so far has focused on the consequences of a doubling of the concentration of carbon dioxide equivalent trace gases from its pre-industrial level.[2] The majority of current General Circulation Models predict

---

[1] This chapter is to some extent based on three papers published in *Energy and Environment* (Zhang, 1994a), *International Journal of Environment and Pollution* (Zhang, 1996b), and in *Intereconomics* (Zhang, 1997d).

[2] This is often labelled as the $2 \times CO_2$ level. $2 \times CO_2$ is a completely arbitrary benchmark, chosen solely for analytical convenience. It is neither a desired situation nor a steady state, and warming will continue (and in fact, will aggra-

7

that an effective doubling of atmospheric CO_2 concentration would lead to an average increase of atmospheric temperature ranging from 1.5°C to 4.5°C (IPCC, 1990). The range of temperature change reflects the uncertainties, particularly with regard to the magnitudes of feedbacks associated with cloud cover, ocean-atmosphere interaction, convection, sea ice and transfer of heat and moisture from the land surface. The uncertainties are even greater in translating such a temperature change into climates and hence in forecasting the resulting consequences of climate change. Present knowledge is able to provide only the basis for developing scenarios that describe the kinds of changes that could occur.

Agriculture is a sector most sensitive to climate change (Nordhaus, 1990, 1991b). In agriculture, any effect depends on the change in temperature, precipitation and soil moisture. Globally, precipitation would rise by 3-15% if there was a doubling of atmospheric CO_2-equivalent concentrations (IPCC, 1990). However, evapotranspiration would rise by the same amount. The net effect would vary between regions. Higher soil moisture is likely only in the high latitudes in winter, where growing seasons will be longer. In contrast, for northern mid-latitude continents (the current grain belts), soil moisture in summer is expected to decrease. The fertilization function of higher levels of CO_2 partially compensates for lower yields in agriculture due to less soil moisture and greater heat stress. Worldwide, if atmospheric CO_2-equivalent concentration doubles, agricultural effects are expected to be negative according to the central estimate by Cline (1992), which shows a loss of some 7% of world agricultural production. The breakdown of regional effects is as follows: average agricultural yields would decline by 20% in the United States and the European Community, 18% in Canada and 10% in China, but rise by 15% in Northern Europe (Cline, 1991; Cline, 1992).[3] Despite the large effects of doubling CO_2 concentration on world agriculture, these effects are generally small when measured as a percentage of national income. As Table 2.1 shows, with the exception of China, no country/region is estimated to experience welfare losses greater than 1% of GDP even under the high impact scenario considered, with the worldwide loss being about 0.5% of world GDP. The relatively small macroeconomic impact is because agriculture accounts for only a small proportion of GDP in most economies - 3% in industrial market economies and 19% in developing economies in 1986 (World Bank, 1988a).

Increasing greenhouse gas concentrations are expected to cause (and have caused) a rise in global-mean sea level, partly due to oceanic thermal expansion and partly due to the melting of land-based ice masses. Sea level rise thus is another source of greenhouse damage. Observations show that the average global sea levels have risen by 10-20 cm over the past 100 years. The IPCC estimates

---

vate) beyond the $2 \times CO_2$ level (cf. Cline, 1992).

[3] Chinese scientists think that the greenhouse effect would reduce China's agricultural output by at least 5% (Xia and Wei, 1994).

that, with business-as-usual, global warming induced by greenhouse gases (GHG) will accelerate this sea level rise, thus leading to an increase of 18 cm by 2030, 44 cm by 2070 and 68 cm by the end of the next century (IPCC, 1990). This IPCC best estimate with business-as-usual represents an implied rate of rise that is about 3-10 times faster than the one experienced over the past 100 years. Even with substantial decreases in the emissions of major GHG, future increases in sea level are inevitable - a sea level rise 'commitment' - due to lags in the climate system (IPCC, 1990).

Table 2.1 *The welfare effects of $2 \times CO_2$ on world agriculture under the high impact scenario (-: losses)*

	Welfare change (millions $ 1986)	As % of 1986 GDP
US	-13027	-0.31
Canada	-738	-0.21
European Community	-13677	-0.40
Rest of Europe	-524	-0.10
Japan	-5614	-0.29
Australia	75	0.04
ex-USSR	-10753	-0.52
China	-12374	-5.48
Brazil	602	0.22
Argentina	2223	2.82
Pakistan	528	1.63
Thailand	490	1.22
Rest of the World	-22513	-0.84
World total	-75302	-0.47

Source: Kane *et al.* (1992).

Increased sea levels have straightforward impacts that include: inundation of low-lying areas and island nations; erosion and recession of sandy shorelines and wetlands; penetration of salt into drinking and agricultural water supplies; damage of infrastructure including harbours, coastal defence systems, roads and other infrastructure. The direct impacts, combined with their resulting socioeconomic consequences, are potentially dramatic, depending on the magnitude and rate of sea level rise. For instance, the IPCC (1991) estimates that a rise of one metre in sea level could inundate 12-15% of Egypt's arable land and 14% of Bangladesh's net cropped area. Worldwide, a rise of one metre would eliminate 3% of the earth's land area, and a larger percentage of its crop areas if no adaptive measures were taken (Rosenberg *et al.*, 1988). This impact will be significant in terms of the cost of protecting and defending coastlines. See the next section for the damage estimates for sea level rise.

Global warming will increase energy demand for space cooling and conversely decrease demand for space heating. Global warming is also expected to put stress on water supply. While some areas may have increased precipitation, water runoff and river flow in many other areas would be likely to decline as a combined result of higher evaporation (due to warmer temperatures) and less precipitation. This will in turn jeopardize energy supply from hydropower generation and biomass, both major energy sources in a large number of developing countries. Rather than by climate change itself, the impact on energy will largely be determined by response strategies against global warming. Changes in patterns of energy generation and usage will have an impact on lifestyles and living patterns, particularly in the developed countries.

Ecosystems on earth are also under threat from GHG-induced climate change. In the case of forestry, the southern boundaries of major forest-type zones and species would shift northwards by 600 to 700 km due to a doubling of CO_2 concentrations (Cline, 1991). Assuming that forests could migrate not faster than 100 km per century (twice the known historic rate), this implies that there would be substantial forest loss and extinction of tree species. Climate change poses also a serious threat to natural terrestrial ecosystems; global biological diversity is expected to decrease and major vegetation zones would face severe disturbances and disruptions. The socioeconomic consequences of these impacts will be significant, particularly for those regions where societies and related economies are dependent on natural ecosystems for their welfare.

Other impacts of global warming include increased morbidity and mortality during summers (with some offset in winters), increased incidence of some vector-borne diseases, increased air pollution, increased destruction from hurricanes, and thawing of permafrost with resulting damage to highways, railways and housing. See IPCC (1990) and Cline (1992) for a detailed discussion.

2.3   Damage estimates for a doubling of CO_2 concentration

If policymakers decide to pursue mitigation policies to slow down climate change, they will probably do so on the basis of anticipated benefits, namely, avoided damages from slowing down the rate of climate change through cutting GHG emissions. This means that, prior to any government action, policymakers need the information on damages caused by climate change.

So far, there have been some studies, though relatively few, that estimate the damages. These studies are based on the benchmark of a doubling of atmospheric CO_2 concentration from its pre-industrial level when estimating the damages. The pioneering study in this tradition is that of Nordhaus. In his study, Nordhaus (1991b) discusses preliminary order-of-magnitude estimates of damages resulting from a doubling of atmospheric CO_2-equivalent concentration. He begins with a breakdown of the US gross national income by sector and subdivides it further into regimes of sensitivity to climate change. As Nordhaus shows about 3% of GNP in the United States originates in climate-sensitive sectors such as agriculture and forestry, and another 10% comes from moderately affected sectors - energy, water systems, property and construction -, while about 87%, the largest share, comes from sectors that are negligibly affected by climate change. The overall impacts of a doubling of CO_2 concentrations on the United States are estimated as follows:

- Damages in agriculture (offset by the CO_2 fertilization effect) are estimated at plus or minus $10 billion as an overall impact on all crops;
- Annual damage from sea level rise is estimated at $5.29 billion, $1.55 billion of which is from loss of land, $0.90 billion from protection of sheltered areas, and $2.84 billion from protection of open coasts;
- Global warming is expected to increase annual electricity demand by $1.65 billion and reduce non-electric space heating by $1.16 billion. A net annual extra cost to the US economy would be $0.49 billion.

Adding the quantified cost items above, Nordhaus estimates that the net annual economic damage resulting from an equilibrium doubling of atmospheric CO_2-equivalent concentrations would amount to $6.23 billion in 1981 US dollars, with 92% of the total damage cost attributed to sea level rise and 8% to energy demand changes. This is equal to 0.26% of annual US national income. Assuming that the composition of world income in 2050 is the same as US income in 1981, Nordhaus extrapolates the US damage 'snapshot' to global annual damages in 2050 on the basis of the scaling ratio of the world income in 2050 to the US income in 1981, and estimates that the annual global economic loss is likely to be around one-quarter percent of total global income. Nordhaus admits that 'inadequately studied or inherently unquantifiable' effects may raise the total damage to 1% of total global income, with an upper bound 'unlikely to be larger than 2% of total output'.

Given the relatively large agricultural sectors in developing countries there, the damages can be expected to be more severe. Thus, many authors have questioned Nordhaus' claim that the estimates based on a developed country can be extended to the world as a whole (cf. Ayres and Walter, 1991; van Ettinger *et al.*, 1991; Grubb, 1993; Fankhauser, 1994). They suggest that the damages due to global climate change may be larger in amounts of dollars in the categories Nordhaus has monetized. Furthermore, the damages may be broader in their scope, potentially affecting numerous other categories that Nordhaus has neither provided dollar estimates for nor considered in a qualitative manner.

A modification of Nordhaus' estimates on damage costs has first been made by Ayres and Walter (1991). With 92% of the total damage costs identified by Nordhaus attributed to sea level rise, Ayres and Walter place emphasis on the implications of this rise. Instead of just extrapolating from the US data to the rest of the world as Nordhaus did, Ayres and Walter estimate the worldwide impacts of sea level rise in terms of assumptions about the lost land and land price; the environmental refugees from the lost land and resettlement costs; and the total coastline protected and the unit cost of protection. Thus, a revised set of economic impacts of increased sea level includes the value of lost land worldwide, the cost of resettling environmental (that is, climate) refugees from the lost land, and the cost of protecting the coastline. Each of the cost items is quantified as follows:

- Cost of protecting 0.5-1.0 million km of coastline: $2.5-5 trillion;
- Cost of loss of 500 million hectares of vulnerable land along 0.5-1.0 million km of coastline: $15 trillion;
- Cost of resettling 100 million of refugees: $1.0 trillion.

Altogether, this adds up to $18.5-21 trillion for the world as a whole over 50 years. Annualized, it comes to about 2.1-2.4% of gross world income. This revised damage cost is nearly 10 times that of Nordhaus' central estimate for total costs, and slightly above his upper bound. Thus, the revised outcomes, though open to question,[4] imply that Nordhaus' analysis may underestimate the potential damages from climate change.

Another direction of improvement of Nordhaus' back-of-the-envelope estimates on damage involves incorporating regional damage. The well-known study in this tradition is that of Fankhauser (1994). To give a flavour of the expected impacts, the detailed results of the Fankhauser study are presented in Table 2.2. This study distinguishes six geopolitical regions and considers twelve damage categories. As Table 2.2 shows, the main sources of damage due to a doubling of CO_2 concentration come from agriculture loss and loss from sea level rise, the latter of which includes coastal protection, dryland loss and wetland loss. Thus it should come as no surprise that the emphasis in the damage discussion has so far been mainly on these two categories. Nevertheless, there are other effects which could be as important and deserve attention. They include impacts on the supply of water, on health, on ecosystems and on the energy sector. With respect to the regional impacts, the results in Table 2.2 support the often expressed view that less developed regions will suffer more than their developed counterparts. Leaving the special case of the former USSR aside, the damage in the non-OECD regions is estimated to amount to 2.2% of their GNP, some 60% higher than the

---

[4] Ayres and Walter use uniform land price for the entitle world. Given that land prices vary a great deal across regions, this treatment is clearly unrealistic. Moreover, by considering both the cost of protecting against sea level rise and the cost of resettling climate refugees from coastal regions, Fankhauser (1994) comes to the conclusion that Ayres and Walter appear to count at least some of the sea level rise impacts twice.

OECD average.[5] As can also be seen, China would be the region hardest hit. As shown in Table 2.2, the highest damage for China is mainly caused by agricultural loss.

Of all the studies discussed above, none of them provide anything better than a rough order of magnitude estimates of potential impacts. The inexact damage estimates are a result of the uncertainties about greenhouse effects and the difficulties in quantifying the damages avoided. Thus, in order to lower the range of error, far more work is urgently required on the magnitude, timing and regional effects of climate change, including the appropriate valuation of economic loss.

2.4   Strategies for responding to climate change

Faced with the likely threat of global warming society is, not surprisingly, being asked to take necessary measures in response. Given the great uncertainties surrounding the likely impacts, however, most countries are pursuing a wait-and-see policy, that is, 'do nothing' and await results of further research. This is likely to endure as long as there is no significant change in scientific consensus or no other political pressures arise relating to the greenhouse effect. Among the countries of this policy stance, we can, in fine tone, distinguish two types of 'do nothing' stance: the one of the North and the one of the South. The former is based on the view that fighting the greenhouse effect is too expensive and useless since the North is less vulnerable to the greenhouse effect, whereas the latter is based on the view that the South is not too much worried about the consequences of the greenhouse effect, and fighting it would unfairly hinder its development (cf. Schelling, 1992; Lipietz, 1995). Whatever 'do nothing' stances are, the downside of these policy stances is that the level of committed warming will have been increased and the damage cost burden may be much higher because of the delayed response (Turner *et al.*, 1994). In contrast, it has been argued that we can probably not afford to wait until all of the uncertainties have been removed. The debate therefore tends to favour 'do something' strategies, because of the potentially serious consequences of climate change.

To date, two main schools of thought have emerged on how to adopt 'do something' strategies: limitation and adaptation strategies (cf. IPCC, 1991).

The first school, with the aim at limiting, stopping or even reducing the growth of GHG concentrations in the atmosphere, has received the greatest public attention. This option is most relevant to policymakers because preventive measures must be taken today (Nordhaus, 1991b). Among the often proposed cheap preventive measures are energy conservation, phase-out of production of the CFCs, reforestation and 'getting prices right' to secure an efficient use of resources. In addition to reducing CO_2 emissions, implementing these measures

---

[5] These figures are calculated based on the numbers given in Table 2.2.

Table 2.2 *Monetary estimates for $2 \times CO_2$ damage across regions (billions of US dollars)[a]*

Type of damage	EU	USA	Ex-USSR	China	Non-OECD	OECD	World
Agriculture	9.7	7.4	6.2	7.8	16.0	23.1	39.1
Forestry	0.1	0.6	0.4	0.0	0.2	1.8	2.0
Fishery[b]	-	-	-	-	-	-	-
Energy	7.0	6.9	-0.7	0.7	2.6	20.5	23.1
Water	14.0	13.7	3.0	1.6	11.9	34.8	46.7
Coastal protection	0.1	0.2	0.0	0.0	0.5	0.5	1.0
Dryland loss	0.3	2.1	1.2	0.0	5.9	8.1	14.0
Wetland loss	4.9	5.6	1.2	0.6	14.7	16.9	31.6
Ecosystems loss	9.8	7.4	2.3	2.2	15.0	25.5	40.5
Health/mortality	13.2	10.0	2.3	2.9	14.8	34.4	49.2
Air pollution	3.5	6.4	2.1	0.2	3.5	11.9	15.4
Migration	1.0	0.5	0.2	0.6	2.3	2.0	4.3
Hurricanes	0.0	0.2	0.0	0.1	1.7	1.0	2.7
TOTAL	63.6	61.0	18.2	16.7	89.1	180.5	269.6
(% GNP)	(1.4)	(1.3)	(0.7)	(4.7)	(1.6)	(1.3)	(1.4)

[a] Negative numbers denote benefits ('negative damage').
[b] Fishery loss is included in wetland loss.

Source: Fankhauser (1994).

would reduce other air pollutants, such as SO_2, NO_x and particulates, and traffic congestion, accidents, road damage and noise. The effects are often called the secondary benefits of carbon abatement (Ayres and Walter, 1991).[6] Thus, proponents of 'no-regrets' policies argue that even if the greenhouse effect turns out to be a non-event, little has been lost and a good deal may have been gained.[7] By contrast, those who do not believe in free lunch take the 'regrets' stance. The 'regrets' policies attack greenhouse gas emissions directly through measures such as emission taxes or tradeable emission permit schemes.[8] They are called 'regrets' policies because the expected costs from emission abatement would not be justified unless there were going to be adverse impacts from the greenhouse effect (Treadwell *et al.*, 1994).[9] Empirical studies show that the 'regrets' policies are very costly; relatively deep cutbacks come at a loss of 1-3% GDP (Cline, 1994). In view of the great uncertainties of climate change, Manne and Richels (1992) propose a 'hedging' strategy that treats the climate change problem as essentially binary - severe or mild - and takes limited action during an initial phase as a 'hedge' until scientific consensus emerges. Their model simulations show that the optimal amount of hedging will depend on the certainty with which we can count on scientific advance and on optimism or pessimism about future non-carbon or carbon-free backstop technologies.[10]

---

[6] A preliminary study by Barker (1993) shows that the secondary benefits of UK carbon abatement would be of the same order of magnitude as the GDP loss from reducing CO_2 emissions. This suggests that, for the UK, air quality improvements appear to be sufficient to justify carbon abatement measures, even excluding greenhouse considerations (Fankhauser, 1994).

[7] In addition to the secondary benefits, the 'no-regrets' actions also offer the prospective extra benefit of learning through experience, that is, of gathering better information about benefits and costs of such actions.

[8] In the next section, these economic instruments will be discussed in great detail with respect to CO_2 emissions.

[9] Chapters 4, 5 and 7 show how the economic impacts of combined 'regrets' and 'no-regrets' policies can be analysed by dynamic optimization models, macroeconomic models and computable general equilibrium models, if the autonomous energy efficiency improvement (AEEI) is used as the proxy instrument for a 'no-regrets' policy and the carbon tax as the instrument for a 'regrets' policy.

[10] Regarding energy supply, Nordhaus (1979) defines a backstop technology as an energy technology based on a resource for which there are no resource constraints. In an economy based on such an energy backstop technology, the economic importance of the scarcity of exhaustible energy resources disappears,

By contrast, the second school of thought favours adopting adaptation strategies, which are to adjust the environment or the ways it is used to reduce the consequences of a changing climate. It is believed that even if a very concerted effort were made today to implement limitation strategies, some adaptation would still be necessary because the climate might already be committed to a certain degree of change given a time lag between GHG emissions and consequent climate change. Moreover, natural climate variability itself necessitates adaptation.

Adaptation would include measures such as strengthening sea defence, changing cropping patterns, organizing population migration and increasing irrigation. It requires that any impact, such as a rise in sea level, should occur sufficiently slowly to allow orderly adjustment to take place. This implicit requirement might leave adaptation strategies potentially expensive and not without risks, either because the measures taken may in the event prove insufficient, or because overestimating future impacts would entail needless expenditures. This, on the other hand, points out that limitation and adaptation strategies should be considered integrated packages complementing each other to minimize net costs (IPCC, 1991). The more GHG emissions are cut through the implementation of limitation strategies, the easier it will be to adapt to climate change.

2.5 Setting targets and the choice of policy instruments for limiting CO_2 emissions

In the previous section, it has been pointed out that limitation strategies are of great policy relevance. If the preventive position were adopted, it seems thus logical to assume that policymakers would wish to pursue such strategies in the most cost-effective manner. How can this goal be achieved? For this purpose, I will discuss the often proposed policy instruments that may be used to control CO_2 emissions, including the command-and-control approach, energy taxes, carbon taxes, tradeable carbon permits and joint implementation.[11] Moreover, in

---

and only capital and labour costs determine energy prices.

[11] The dominance of CO_2 contribution to global warming suggests that CO_2 must be the main target in any attempt to limit GHG emissions. For this reason, I limit myself to the problems of controlling CO_2 emissions when discussing policy instruments. But it should be emphasized that including not just CO_2 but also other greenhouse gases will induce more effective options for greenhouse gas control, but certainly complicate the discussion. In that case, instead of CO_2 emissions, a multiplication index such as the Global Warming Potential as proposed by the IPCC (Houghton *et al.*, 1990) would have to be used. See, for example, OECD (1991, 1992) for a further discussion.

discussing these policy instruments, special attention is paid to the economic instruments.

The rest of the section proceeds as follows. Section 2.5.1 points out why the targets for emission reductions need to be predefined. In Section 2.5.2, the differences between energy taxes and carbon taxes are discussed briefly and some main findings arising from those studies on carbon taxes are presented. Section 2.5.3 focuses on some aspects of domestic carbon tax design and incidence. The allocations of emission permits (or reimbursement of carbon tax revenues) are the subject of Section 2.5.4. Section 2.5.5 gives a brief comparison of carbon taxes with tradeable carbon permits. Section 2.5.6 discusses joint implementation from a Chinese perspective.

2.5.1 Setting targets: a separate approach

Before dealing with policy instruments targeted at the control of CO_2 emissions, I will discuss briefly why targets for emission reductions need to be predefined.

According to the conventional theory of environmental economics, there is no need to set the targets for emission reductions beforehand when the associated externalities are internalized (cf. Baumol and Oates, 1988). The optimal emission level is achieved when the point is reached at which the marginal cost of reducing emissions is the same as the marginal cost of the damages. As long as the so-called Pigouvian tax is set equal to the marginal cost of the damages, its implementation automatically leads to the optimal situation. This means that the process of internalization itself co-determines the target. However, this principle works better for conventional environmental problems than for those problems with international and intertemporal dimensions (for example, acid rain, ozone layer depletion and climate change), an essential feature of which is the absence of an institution with the international jurisdiction to enforce policy (cf. Folmer *et al.*, 1993). This also has consequences for the formulation of policy, including the revelation of costs and benefits (cf. Barrett, 1990).

Given the characteristics of these problems, we need to reconsider a separate approach that was first proposed by Baumol and Oates (1971; also 1988): setting emission reduction targets first and then selecting instruments to achieve these targets at the least cost. Compared with the conventional approach where the optimal solution is sought, effectiveness is the goal for the Baumol-Oates case in which there is nothing indicating that the level of emission reduction achieved by the separate approach is either the economic or even the environmental optimum (cf. Baumol and Oates, 1971 and 1988; Faucheux and Noel, 1991).

It has been observed that the two most important international agreements on limiting emissions of atmospheric pollutants - the 1987 Montreal Protocol on Substances that Deplete the Ozone Layer and European Community's Large Combustion Plant (LCP) Directive to limit acid emissions - have been formulated in this way. The Montreal Protocol, in its original form, calls for a 50% reduction in CFC emissions by the signatory countries by 1999, with a grace period of ten years for developing countries (cf. Enders and Porges, 1992).[12] The LCP Directive incorporates a complex formulation of SO_2 and NO_x reduction levels for three target dates, with different elements of backdating for each member country (cf. Haigh, 1989).

With the conclusion of these two major agreements based on percentage reduction targets for gaseous emissions, it is not surprising that calls for limiting CO_2 emissions have focused on a similar strategy. The Toronto Conference has recommended a 20% reduction by 2005 and a 50% reduction by 2025 in global CO_2 emissions relative to the 1988 levels, with an initial goal set for a 20% cut by 2005 in the industrialized countries (UNEP, 1988).

The acceptable reduction targets can be set by scientific expertise or international agreement. Whatever the acceptable carbon reduction target that is eventually set, the remaining issue is how it is to be achieved. In this regard, there are five alternative policy instruments: the command-and-control approach;[13] energy taxes; carbon taxes; tradeable carbon permits; and joint implementation.

With regard to the global warming problems, especially in the CO_2 context, a number of recent studies discuss market-based instruments or economic incentive instruments, namely energy taxes, carbon taxes, tradeable carbon emission permits and joint implementation. It is argued that these economic instruments to limit CO_2 emissions can achieve the same target at lower costs as compared with the conventional command-and-control regulations. Moreover, the economic instruments can act as a continuous incentive to search for a cleaner technology, while, for the command-and-control regulations, there is no incentive for the polluters to go beyond the standards, unless the standards are continually revised and set slightly above the best available technologies (Tietenberg, 1992). Therefore, the economic instruments have a *technology-forcing* characteristic. Some evidence shows that this dynamic efficiency aspect of economic instruments is important (Tietenberg, 1990). In the CO_2 context, the dynamic efficiency takes on extra dimension because, unlike sulphur, CO_2 is difficult to dispose of, even if it is removed from stack gases, and incentives to develop disposal technologies are therefore of particular relevance (Pearce, 1991).

In the following sections, the discussion is restricted to the economic instruments in the CO_2 context, namely energy taxes, carbon taxes, tradeable carbon emission permits and joint implementation.

---

[12] The Protocol has since been amended and strengthened in a number of aspects.

[13] In the international CO_2 context, this approach includes the widely-discussed uniform percentage reductions in emissions by all participating countries. In this case, individual countries would be left to reduce their CO_2 emissions by, for example, traditional command-and-control regulations, tradeable carbon emission permits for domestic sources, or domestic carbon taxes.

2.5.2 Energy taxes versus carbon taxes

An energy tax is an excise tax, which is defined as a fixed absolute amount of, for example, US$ per Terajoule. It is a tax imposed on both fossil fuels and carbon free energy sources according to their energy (or heat) contents, with renewables usually being exempt. By contrast, a carbon tax (an excise tax that is imposed according to the carbon content of fossil fuels) is restricted to carbon-based fuels. Given that oil and gas have greater heat contents for a given amount of CO_2 emissions as compared with coal, an energy tax lies more heavily on oil and gas than a carbon tax. Moreover, an energy tax burdens nuclear energy, which, with the exception of hydropower, provides the only so far proven method for enormous potential for the large-scale generation of electricity without a directly parallel production of CO_2 emissions.

If the goal is to reduce CO_2 emissions, a carbon tax is preferred on grounds of cost-effectiveness, given that a carbon tax is able to equalize the marginal cost of CO_2 abatement across fuels and therefore satisfies the condition for minimizing the cost of reducing CO_2 emissions. This implies that an energy tax (if introduced) will lead to poor target achievement or else to unnecessarily high costs as compared with a carbon tax (cf. Kågeson, 1991; Cline, 1992; Manne and Richels, 1993; Jorgenson and Wilcoxen, 1993b). This can be explained by two factors: price-induced energy conservation and fuel switching (Manne and Richels, 1993). Carbon taxes reduce CO_2 emissions through both their price mechanism effects on energy consumption and fuel choice. By contrast, since the energy tax is imposed on both fossil fuels and nuclear energy, the incentive for fuel switching will be reduced, and the reductions in CO_2 emissions will be mainly achieved by price-induced energy conservation. Thus a higher tax is required for achieving the same reduction target as compared with the carbon taxes. Put another way, for the economy in question it is more costly to reduce CO_2 emissions through an energy tax than through a carbon tax. This has clearly been shown by the study of Manne and Richels (1993), which evaluates the implications of the CEC proposal for a mixed carbon and energy tax.[14] Similar findings are also shown by the studies of Jorgenson and Wilcoxen (1993b) and Beauséjour *et al.* (1995). The results of Jorgenson and Wilcoxen suggest that in

---

[14] Recognizing that a carbon tax puts a relatively high pressure on coal, the most secure energy supply, and that both a carbon tax and an energy tax have a quite different impact on member states, a carbon/energy tax has been proposed by the Commission of the European Communities (CEC) as part of its comprehensive strategy to control CO_2 emissions and increase energy efficiency. The CEC proposal is that member states introduce a carbon/energy tax of US$ 3 per barrel oil equivalent in 1993, rising in real terms by US$ 1 a year to US$ 10 per barrel in 2000. After the year 2000 the tax rate will remain at US$ 10 per barrel at 1993 prices. The tax rates are allocated across fuels, with 50% based on carbon content and 50% on energy content (cf. CEC, 1991).

2020 the US GNP loss from an energy tax is 20% greater than that from a carbon tax in order to stabilize the US CO_2 emissions at 1990 levels in the year 2020. The results of Beauséjour *et al.* indicate that in 2000 Canada's GDP loss from an energy tax is 20% greater than that from a carbon tax in order to stabilize Canada's CO_2 emissions at 1990 levels in the year 2000. While being the more cost-effective of the tax instruments considered, the carbon tax is also less burdensome in that it raises a smaller amount of government revenues for a given reduction of CO_2 emissions (Jorgenson and Wilcoxen, 1993b; Beauséjour *et al.*, 1995).

Let us now turn to the carbon tax. So far, a number of studies have focused on the cost estimates for achieving a given reduction in CO_2 emissions. These studies usually incorporate a carbon tax as a method to achieve the target because of its effectiveness. The main findings arising from these studies (cf. Whalley, 1991; Whalley and Wigle, 1991a, 1991b; Martin *et al.*, 1992; Hoeller and Coppel, 1992; Piggott *et al.*, 1992; Pezzey, 1992; Walker and Birol, 1992; Manne and Richels, 1991a, 1993; Felder and Rutherford, 1993; Kverndokk, 1993; Jorgenson and Wilcoxen, 1993a, 1993b; Martins *et al.*, 1993; OECD, 1993b; Poterba, 1993; Manne, 1994) are that, among other things,

- the carbon tax should increase over time if it is to reflect the rising cost of damage from the accumulation of CO_2 concentration in the atmosphere, if it is to give the markets the signal that CO_2 emissions will eventually be heavily taxed, and if there are few economically feasible substitutes available;
- there would be significant variation in timing and size of the carbon taxes among countries and regions, given that the marginal cost of abating CO_2 emissions substantially differs across countries and over time;
- the carbon tax could be production- or consumption-based, but the effects across options would be significantly different among countries. A national production-based carbon tax operates much like an export tax. If applied, oil-exporting countries such as OPEC would gain substantially, but in the case of a national consumption-based tax, they would suffer considerably;
- the carbon taxes imposed unilaterally or even regionally would be largely ineffective. This ineffectiveness is attributed partly to a relatively small share of the coalition (for example, EU, OECD) emissions in the world total and partly to strong economic growth and the resulting increase in emissions taking place in non-coalition countries that offset the coalition's achievements;[15]
- the autonomous (that is, non-price-induced) energy efficiency improvement, the possibilities for fuel substitution, and the availability of backstop technologies are essential. Without non-fossil fuel options, the upper bound on the required

---

[15] This phenomenon is the so-called carbon leakage, with its average leakage rate being defined as the ratio of carbon emission *increase* outside the coalition to carbon emission *cutbacks* within the coalition relative to their reference levels (cf. Felder and Rutherford, 1993).

carbon tax would rise. Moreover, the autonomous energy efficiency improvement, and the cost and availability of low-carbon or carbon-free backstop technologies are crucial to limiting the tax level required[16] and thus reducing the costs incurred for compliance with emission reduction targets; and

- the carbon tax itself would impose a deadweight loss on a country where there are no distortions in the energy markets. But when existing distortions arising from energy subsidies are taken into account or when the revenues generated from the imposition of a carbon tax are recycled to the economy for replacing another indirect tax, the introduction of a carbon tax could even lead to a net gain.[17]

I will not go into these interesting topics any further, but instead focus on three aspects that are considered important when designing a domestic carbon tax.

2.5.3 Three aspects of domestic carbon tax design and incidence

The three aspects to be addressed are: (i) the treatment of the carbon tax revenues, (ii) the impacts on the distribution of income, and (iii) the effects on international competitiveness.

I begin with the treatment of the carbon tax revenues. It has been argued that there is a 'double dividend' from the carbon tax (Pearce, 1991):[18] not only environmental dividend through reduced emissions of pollutants but also non-environmental dividend in terms of a reduction in the overall economic cost of

---

[16] As assumed in the GREEN model, the backstop technologies are produced at a constant marginal cost, without any constraint on supply (cf. Burniaux *et al.*, 1992). Thus, the carbon tax needs not to increase further. If there are few economically feasible substitutes available, however, the effectiveness of a carbon tax is likely to be much more limited. Thus, to lower CO_2 emissions very substantially would require a high carbon tax - certainly higher than the taxes already imposed (Barrett, 1991).

[17] For example, the results based on the GREEN model clearly indicate the net gains for the Eastern Europe and former Soviet Union if the existing energy subsidies are taken into account.

[18] In the study of Goulder (1994), the *weak* 'double dividend' and the *strong* 'double dividend' are distinguished. The *weak* 'double dividend' proposition states that in welfare terms the non-environmental dividend is always positive as a reduction in distortionary taxes is always superior to a reduction of lump-sum taxes. In the *strong* 'double dividend' proposition, it is stated that the non-environmental dividend is larger than the gross costs. If the strong claim held, it would reduce the amount of information that policy analysts need to make a benefit-cost case for green tax swaps. See Goulder (1994) for further discussion.

raising government revenues (Lee and Misiolek, 1986).[19] This 'double dividend' feature of a carbon tax has important implications for 'green tax swaps' for distortionary taxes, because different taxes have different distortionary effects on the economy. If the objective of a carbon tax is to reduce consumption of carbon-based energy products through reallocating spending away from CO_2-emitting activities and thus slow down (or even stabilize) the build-up of atmospheric CO_2 concentration rather than for macroeconomic management, the carbon tax is in essence an incentive tax rather than a revenue raising tax. In macroeconomic terms it seems therefore appropriate that revenues raised through an increase in one indirect tax (a carbon tax) could be offset by a reduction of another indirect tax, for example, value added tax (VAT) so as to minimize the effect on the general level of prices.[20] This has been confirmed by the studies of DRI (1991), Standaert (1992), Walker and Birol (1992), and Barker *et al.* (1993), the results of which show that reducing VAT offsets the carbon tax's inflation more than reducing other taxes. The studies of Karadeloglou (1992) and Standaert (1992) also show that the effects in the case of reducing VAT on both GDP and employment are less negative than those in other tax offset cases (see Table 4.2).[21] Another measure used to recycle all revenues from the carbon tax to the economy is by means of reducing income tax. If this is adopted, inflation is likely to increase, although the extent of acceleration depends on the character of wage negotiations for increases in disposable income resulting from the reduction in income tax. This higher inflationary response has been found in the modelling of the effects of the CEC tax (cf. DRI, 1991; Karadeloglou, 1992; Standaert, 1992; Barker *et al.*, 1993). Alternatively, if the carbon tax revenues are retained in treasury coffers to reduce public sector deficits, then this will depress the economy, certainly in the short term. If the revenues are all spent by the government, for example on non-fossil energy investment, this would imply a large

---

[19] The non-environmental dividend is very often interpreted as using the extra carbon tax revenues to reduce existing distortionary taxes for raising government revenues. This dividend can of course have other interpretations. In the study of Bovenberg (1994), for instance, reduced unemployment is referred to as the potential extra dividend in addition to improved environmental quality. In the context of tradeable carbon permits, the extra dividend refers to the proceeds from the sale of carbon emission permits (Manne and Richels, 1995).

[20] A carbon tax, by raising the prices of fossil fuels, will raise the general level of prices. Offsetting it with reductions in VAT or other taxes tends to lower the price level, but the price effect is expected to vary, depending on the tax offset arrangements.

[21] See Section 7.4 for analysing the economic impact of carbon taxes for China under the four indirect tax offset scenarios.

investment programme which could lead to macroeconomic imbalance and rapid inflation (Barker *et al.*, 1993).

The second aspect of a domestic carbon tax is its impact on the distribution of income. A carbon tax would have a regressive impact on the distribution of income since lower income households spend a larger proportion of their income on energy than higher income households. Smith (1992) calculates the distributional effects of a mixed carbon and energy tax at $10 per barrel in the UK on different income groups. The results show that the poorest 20% of the population would have to pay an additional tax of £1.45 per week, the richest 20% an additional £2.95 per week, and the average household an additional £2.21 per week. Translated into increases of tax paid as a percentage of total spending, these figures are equivalent to 2.4%, 0.8% and 1.4% respectively. Clearly the relative burden of the additional tax would be heavier for the poorest decile, and lower for the richest. This highlights the fact that unless low income groups are to be made worse off by the carbon tax, a large part of the revenues from the tax will need to be used to compensate poorer households suffering from the tax, through tax reductions and increases in social security benefits and pensions to provide roughly uniform amounts in compensation for each household. Unfortunately, using the carbon tax revenues in this way will reduce the scope for the revenues to be used to maximize the efficiency gains from reductions in other existing distortionary taxes, for example, VAT described above (cf. Smith, 1992; Barker, 1992; Pearce, 1991). Thus there is a clear trade-off between efficiency and equity in the use of the revenues: the efficiency gains can only be achieved by sacrificing the distributional neutrality of the package (Smith, 1992).

The findings above are typically shown in the studies for the industrialized countries. Shah and Larsen (1992) argue that such findings cannot be generalized for the developing countries, where the incidence of carbon taxes would be affected by institutional factors. Among some important factors that may have a bearing on the tax-shifting are market power, price controls, import quotas, rationed foreign exchange, the presence of black markets and tax evasion and urban-rural migration (see Shah and Larsen (1992) for a further discussion).

Now I will consider the third and last aspect, namely the effects on international competitiveness. A domestic carbon tax has important implications for the international competitiveness of economies in relative terms. Although international competitiveness is not necessarily reduced over the long term by higher energy prices, in certain industries, the effects of introducing a unilateral carbon tax may be serious in the short term. Exemptions from the new taxes are therefore suggested to protect the price competitiveness of these industries in international trade. For example, the CEC proposal (CEC, 1991) provides for exemptions for the six energy-intensive industries, such as iron and steel, non-ferrous metals, chemicals, cement, glass, and pulp and paper. As discussed earlier, a carbon tax is intended to fall most heavily on the products of carbon-intensive industries. Clearly, the exclusion of these industries from coverage of the carbon tax on grounds of competitiveness reduces the effectiveness of the carbon tax in achieving its objective of reducing CO_2 emissions,[22] while it does mean that the EC industries most vulnerable to competition are protected in their markets. The ineffectiveness of the EC unilateral action suggests that at least similar actions in competitor countries, especially in the United States and Japan, should be taken (or some more general OECD-wide tax should be adopted), although carbon taxes need to be imposed globally in order to achieve sufficient reductions in CO_2 emissions.

So far this discussion has been restricted to domestic carbon tax. It has been argued that even if domestic emission reduction targets are achieved in cost-efficient ways, for example through a domestic carbon tax, a global cost-efficient emission reduction target can only be achieved if CO_2 emissions are distributed among countries in such a way that the marginal cost of abatement is equalized among countries (cf. Hoel, 1991, 1992). This global cost efficiency may be achieved through either an international carbon tax or a tradeable carbon emission permits regime. The remainder of this section will deal briefly with the former, while the latter will be left to be discussed in the next section.

Hoel (1991) has shown that a tax administered and collected by an international agency is too bureaucratic and would interfere with domestic sovereignty, while a tax implemented by each government would fall foul of free rider problems, since governments could easily offset a carbon tax by reducing other domestic taxes on fossil fuels. Therefore, the way out has to be one in which the carbon tax should be globally imposed on each country by some international agency but nationally administered and collected through its central government (cf. Hoel, 1991). The carbon tax is set to be the same for each country. The revenues from the tax are then reimbursed; handed back to the countries where the revenues are raised according to some agreed rule of allocation. Each country would then act to minimize the sum of its tax payments and abatement costs. How, then, should the tax revenues be reimbursed? This is equivalent to the determination of the initial permits under a regime of tradeable carbon emission permits.[23] The issue is to be discussed in the next two sections.

---

[22] In addition to this limitation, there are two more problems. The first is that the industries which are exempt from paying the CEC tax will improve their competitive position in relation to those industries which are not. There will therefore be some switching of demand towards the products of these energy-intensive industries, which is precisely the reaction that such a tax should avoid. The other problem is that firms which find themselves paying the tax will try to be reclassified as exempt or eligible for rebates if at all possible, thus limiting the impact of the tax on energy consumption and CO_2 emissions (cf. Barker *et al.*, 1993).

[23] A carbon tax regime, in which total CO_2 emissions are equal to X and tax reimbursements to the n participating countries are proportional to the vector $(a_1, a_2,...,a_n)$ with $\Sigma_i\ a_i = 1$, is equivalent to a regime of tradeable carbon permits, in

2.5.4 Tradeable carbon emission permits

An alternative to an international carbon tax is a regime of tradeable carbon permits, which allows the permit holders to trade or sell their entitlements to other countries (Pearce, 1990; Hoel, 1991). As long as the marginal cost of reducing CO_2 emissions differs among countries, countries have an incentive to trade permits with the market price of CO_2 permits being equal to the marginal cost of abatement, and make net gains. The process continues until the marginal cost of reducing CO_2 emissions is just equalized across countries, thus inducing a cost-efficient distribution of CO_2 emissions.

Once an international emission budget is set, the question then arises how to allocate the initial emission permits to each participating country.[24] The obvious rules are based on both the cost of reducing CO_2 emissions and the consequences of climate change. The rules could be applied if the cost of abatement and the consequences of climate change were common knowledge (Hoel, 1991). However, this is not the case.[25] In practice, these costs cannot be measured objectively with any precision, and there are still uncertainties regarding the magnitude, timing and regional effects of climate change. For this reason, the allocation of permits would in practice have to be based on relatively straightforward rules. In the CO_2 context, the rules based on uniform percentage reduction, historical CO_2 emissions (a grandfathering approach), current GNP (or GDP), and population, among other rules, have been suggested (cf. Grubb, 1989;

---

which the initial permits allocated to the n participating countries are $(a_1X, a_2X,...,a_nX)$. See Hoel (1991) for further discussion.

[24] *Ceteris Paribus* entitlements defined in terms of emissions would be preferred; this would produce the most cost-effective outcome. With an emissions target, the tendency of the market to seek the least cost means of control would be focused on reducing emissions, which, of course, is the objective (see Bohm (1993) for further discussion).

[25] The cost of reducing CO_2 emissions differs significantly across countries, depending on among other factors economic structure, products mix, fuel mix, current efficiency of energy use, current levels of energy price, and availability of backstop technologies. Moreover, given the huge uncertainties surrounding the magnitude, timing and regional effects of climate change, any estimate for the consequences of climate change for different countries must be speculative. Thus, if the allocation of initial permits were based on the rules, it would be in each country's interest to claim that it found reducing CO_2 emissions burdensome and climate change not very harmful. The negotiation process would be extremely difficult and it is doubtful whether any agreement would be reached (Hoel, 1991). This does not of course mean that there would be no difficulty of allocating the initial permits according to other rules, which are discussed below.

Pearce, 1990; Rose, 1990; Manne and Richels, 1991b; Hoel, 1991; Cline, 1992; Kverndokk, 1993; Welsch, 1993; Rose and Stevens, 1993; Larsen and Shah, 1994). The diversity of these allocation rules, each of which is discussed below, reflects the lack of consensus on a 'best' equity principle.

A uniform percentage reduction offers the operational advantage because it focuses on easily observable physical burden-sharing (Welsch, 1993). It is for this reason that international environmental agreements often take the form of a uniform percentage reduction. An example is the 1985 Helsinki Protocol on the Reduction of Sulphur Emissions or Their Transboundary Fluxes by at least 30% (cf. Shaw, 1993).[26] In the CO_2 context, the rule ignoring the past build-up and simply basing reduction requirements on current emissions would be equivalent to penalizing developing countries for their economic development when no such penalty was imposed on industrialized countries for their abusing of the global commons in the course of their industrialization (Rose and Stevens, 1993). Therefore it at least seems conceivable that the rule would not be accepted by the developing countries. Moreover, it has been argued that the rule based on uniform percentage reductions is inefficient in the sense that the same goal could be achieved at lower costs through the rule that equalizes the marginal cost of abatement among all participating countries (cf. Hoel, 1992).

Using the grandfathering of permits, which is based on past CO_2 emission levels, or current GNP as a base would minimize the disruption of current production. However, using either rule as a base would favour the developed countries and does little or nothing to create incentives for the developing countries to cooperate. Moreover, there are some nasty wrinkles associated with adjusting the initial permits: Should an energy-efficient country such as Japan be rewarded with additional permits? Should a country that relies on nuclear power and therefore is a small emitter such as France get extra permits? Should Brazil,

---

[26] Because of great concern about the long-range transboundary flow of sulphur and nitrogen oxides and the resulting regional-scale environmental damages (such as acidification of soil and fresh water and damage to vegetation), at a ministers' meeting in Ottawa, Canada, in March 1984, 10 nations volunteered to reduce emissions of sulphur dioxide by 30% by 1993 relative to their 1980 levels. This group of nations was referred to unofficially as the '30 Percent Club'. By June 1984 membership in the '30 Percent Club' had increased to 18 nations. In July 1985 in Helsinki, Finland, a protocol to reduce sulphur dioxide emissions or transboundary fluxes by at least 30% was signed by 21 nations. Among the nations that did not sign the protocol were two of Europe's largest emitters: the United Kingdom and Poland. The former did not sign because, in its opinion, insufficient credit was given in the protocol for past emission reductions, because of the arbitrary choice of both the base year (1980) and the percentage of emission reduction, and because of the lack of connection of deposition levels with environmental impact, whereas the latter did not sign because of its lack of technologies and equipment to control sulphur emissions (cf. Shaw, 1993).

whose copious forests absorb carbon dioxide, be rewarded for that (Sun, 1990)? Also, should countries that have unilaterally cut down their CO_2 emissions long before any CO_2 agreement be rewarded for that, which would have otherwise been punished?

Using population as a base is compatible with equal emission rights and could be accepted as fair by the developing countries. Given the great disparities in current per capita CO_2 emissions, however, this would probably imply the net payments transfers from the developed countries to the developing countries on a substantial scale and therefore would not be easy for political leaders to justify. The study of Kverndokk (1993) shows that transfers are required of 6% and 3% of their potential GDP in the year 2000 from the USA and other OECD countries to the developing countries respectively. The magnitude of these transfers is scarcely credible, considering the United Nations' level of development assistance at 0.7% of GDP has not been met by most of the industrialized countries yet. Moreover, as pointed out by Grubb (1989), the allocation rule may create an implicit incentive for countries to increase their population,[27] whereas just the opposite is needed to address the greenhouse problem. Grubb suggests that only adults above a specific age should be accounted in order to avoid the implicit reward for overpopulation.

The foregoing discussion clearly indicates that the acceptability of tradeable permit regimes will depend on the allocation rules for permits. In view of the respective weaknesses of each rule discussed above, it follows that an acceptable allocation rule should take into account historical CO_2 emissions, GNP (or GDP) and population together, and that the emissions entitlements to each participating country should be adjusted over time in order to reduce the relative benefits and relative excess costs for each country. Pearce (1990), for instance, argues that an allocation regime, based initially on grandfathering but with the emission permits being modified by altering the value of the permits over time, would be most appropriate. Thus, developed countries would have declining permits over time, while developing countries could have rising permits that less than offset the developed countries' reductions. This can be illustrated by, for example, the following formula (Cline, 1992):

$$Q_i = Q^g[w_H\ \Phi_{O,H,i} + w_Y\ \Phi_{O,Y,i} + w_P\ \Phi_{O,P,i}]$$

where i represents the country in question; Q is the emissions quota; superscript g the global emissions target; subscripts H, Y, and P refer to historical CO_2 emissions, GDP, and population respectively; w refers to the weight assigned to the rule, with the sum of w_H, w_Y and w_P being equal to one; Φ is the country's

---

[27] Given that there are many other problems associated with population growth and that the governments are concerned with per capita income (or per capita welfare), for countries to increase their population for this reason is highly unlikely. It is nevertheless worthwhile pointing out this possibility.

share in the relevant global total; and subscript O refers to the base year. This approach weights three alternative rules to determine an overall country permit. Cline argues that, if the three weights shifted over time towards the population rule and thus equity, the approach would seem to stand the best chance of support by both industrialized and developing countries: it would give heavy weight to the 'realism' concerns of industrialized countries at the beginning of the period, but also provide large scope for a shift over time towards the equity concerns of developing countries.

Quantitative analyses of the effects of changing CO_2 permits over time along this line have been made, although relatively seldom. In the study of Manne and Richels (1991b), for example, the carbon permits, though still benchmarked against 1990 base year, are distributed with grandfathering initially (the year 2000) but in proportion to the 1990 level of population at the end of the planning horizon (the year 2100). This allocation rule is designed to not only avoid an abrupt change in the status quo, but in the long run lead to an egalitarian distribution of carbon permits. The results (Manne and Richels, 1991b) show that according to the allocation rule there are no dramatic gains from trade and each of the regions would *benefit modestly from trade*, since none of the five regions buys or sells more than 5% of the total global volume of tradeable carbon permits.

So far this discussion on the allocation rule has been associated with a regime of tradeable carbon permits. These rules are also valid for determining how to reimburse carbon tax revenues if the international carbon tax is imposed across countries.

2.5.5 Carbon taxes or tradeable carbon permits?

As discussed earlier, both carbon taxes and tradeable permits minimize overall abatement costs by allocating the cutbacks to the countries where marginal cost of reducing CO_2 emissions is the lowest. Moreover, given both perfectly competitive markets and certainty, carbon taxes are equivalent to tradeable permits (cf. Hoel, 1991). In practice, however, there are some differences between these two instruments.

Probably the most valid arguments in favour of tradeable permits rather than taxes so far are as follows:

Tradeable carbon permits, unlike carbon taxes, are a form of rationing and the great advantage is that in this way one can be sure of achieving the target agreed. By contrast, the actual achievements in reductions of CO_2 emissions by a proposed carbon tax remain uncertain because of imperfect knowledge of the price elasticities of demand and supply for fossil fuels, especially for the large price increases caused by carbon taxes for major emissions cutbacks - see, for example, Cline (1992) for further discussion. This implies that setting the initial tax will be a hit-and-miss affair, and could thus induce hostile reactions from countries, industries, and consumers although it is not clear how serious an objection this is (Pearce, 1990, 1991). Moreover, in the context of global warming, the delays in adjusting the insufficient carbon tax to the desired level will mean additional committed warming.

Another complication of the carbon tax is the initial difference in energy prices. As a consequence of existing distortions by price regulations, taxation, national monopolies, barriers to trade and so on, there are initially great differences in energy prices, both between fuels and across countries (cf. Hoeller and Coppel, 1992; Haugland *et al.*, 1992). If CO_2 emissions are then to be reduced by similar amounts in two countries, *ceteris paribus*, lower taxes are required for the country with low prices before the tax imposition than for the country with the higher pre-tax prices (cf. Hoeller and Coppel, 1992). Thus an eventual cost-efficient regime of international carbon tax would presumably need to remove existing distortions in international energy markets. Otherwise, countries with the lower pre-tax prices would enjoy free-rider benefits (Cline, 1992).

However, a regime of tradeable carbon permits is also subject to important limitations. In practice, the regime requires a sufficient number of traders (or participating countries) to avoid ill-functioning permit markets,[28] although this will increase its administrative costs. This requirement heightens the importance of large participation of developing countries to avoid an insufficient number of traders. As discussed earlier, however, because of great difficulty in allocating initial carbon permits, it may take quite a long time to induce developing countries to join.[29] Moreover, even if the regime was put into operation, which might require less of an international bureaucracy than would be needed to administer and enforce an international carbon tax, some supra-national agency would be required:

---

[28] When permit markets are thin, that is, when permits are infrequently traded, clear price signals are absent, thereby impairing the functioning of the permit regime. Thus, an international market seems a minimum requirement. Carbon emission permits, traded internationally, allow marginal cost of abatement to be equalized across countries. Permits may be traded independently within nations so that marginal cost of abatement is equalized across domestic sources (Shah and Larsen, 1992).

[29] The 1987 Montreal Protocol on CFCs can be taken as an example. For CFCs, (1) there exists substantial scientific evidence that CFCs play the greatest role in depleting the stratospheric ozone layer; (2) the number of key countries involved in the global production of CFCs, the overall economic cost of phasing out CFCs, and institutional changes involved are relatively small; and (3) the oligopolistic nature of the CFC-producing industry ensures that producers' cooperation could be secured by effective cartelization and limitation of production, making the monitoring of compliance not too difficult (Enders and Porges, 1992; Shaw, 1993). Even for this case, which is far less complicated and costly than that of greenhouse emissions, it still took over ten years to reach the Protocol.

1. to regulate and perhaps periodically intervene in the permit market in which some undesirable consequences may occur. Hoel (1991), for instance, argues that big countries can influence prices for permits. For a larger seller, it is optimal to have higher carbon emissions than the level indicated by the marginal cost of abatement (that is, the market price for permits). The opposite holds true for a larger buyer;[30]

2. to adjust the global target level and re-issue permits in response to changing conditions as discussed earlier; and

3. to monitor transactions and enforce any penalties for abuse (cf. Pearce, 1990; Hoel, 1991; Pearce and Barbier, 1991).

All the administrative and transaction costs associated with tradeable permits cannot be known in advance. They may turn out to be much higher than was imagined when the target was defined, thereby making tradeable permits less of an attractive instrument. This uncertainty regarding the cost of emission reductions is an important distinction between tradeable permits and carbon taxes. Weitzman (1974) has shown that, under specified conditions, if the marginal abatement cost curve is steeper than the marginal damage curve for emissions, then the cost of making an error in the selection of a price-based instrument such as an emission fee or charge will be less than that of making an error in the selection of a quantity-based instrument such as a tradeable permit. This suggests that if there is great uncertainty about the cost of emission reductions, carbon taxes are preferred in order to avoid potentially large and unexpected costs (Shah and Larsen, 1992). However, if the overall impacts of climate change are believed to be unacceptably high or if there was a threshold effect caused by the stock of CO_2 emissions beyond which atmospheric temperatures would rise exponentially, the target would then have high political priority. In this case, the choice of economic instruments should not be swayed by uncertainty regarding the cost of emission reductions and tradeable permits would be preferred to carbon taxes (cf. Kågeson, 1991; Shah and Larsen, 1992).

Moreover, so far there has been limited international experience with tradeable permits.[31] While tradeable permits have enjoyed some considerable success in the various domestic contexts, this by no means guarantees their success in international context (Tietenberg and Victor, 1994). Thus such a regime should perhaps be validated through more experience on a small rather than a global scale. In this regard, it is worthwhile putting into practice joint implementation, a

---

[30] See Hahn (1984) and Misiolek and Elder (1989) for a theoretical analysis of tradeable emission permits when some of the participants have market power.

[31] There are two main applications of tradeable permits - emissions trading under the US Clean Air Act (Tietenberg, 1990) and the use of individual quotas as the primary means of addressing overfishing and depletion of inshore stocks in New Zealand and other countries (OECD, 1993b).

derivative of the idea of permits trading. The experiments may provide some experience for implementation of a global tradeable emission permits regime (cf. IEA, 1992; Tietenberg and Victor, 1994).[32] This brings up the issue of joint implementation.

2.5.6 Joint implementation

In 1992, the Norwegian delegation introduced the concept of joint implementation (JI) into the negotiations for the United Nations Framework Convention on Climate Change (FCCC, hereafter also referred to as Climate Convention) aimed at, in the long term, stabilizing greenhouse gas (GHG) concentrations in the atmosphere. At the Rio Conference on Environment and Development, JI was put into the final text of Article 4.2 of the FCCC that 154 countries and the European Union signed. This is deemed a breakthrough for JI as a climate policy instrument. The inclusion of JI in the Climate Convention can also be regarded as a first step towards a global regime of tradeable emission permits.

In brief, JI means that the investor country invests in emission abatement projects in another (host) country where the costs of abating GHG emissions are lower than trying to achieve an equivalent abatement within the own country and is credited, in whole or in part, for emission abatements in its own GHG accounts. JI enables the investor countries to 'shop around' for the lowest way to limit emissions. Thus, it offers potential for reducing the global costs of GHG abatement. This is the economic rationale for JI.

Then, how should JI be implemented? Because a number of countries have been sceptical about JI during the negotiations for the Climate Convention, the Convention therefore offers no specific guidance on the application of JI and leaves it to the Conference of the Parties (CoP) to lay down the rules. Now, as the pilot JI projects are being launched, attention is increasingly focused on the actual implementation of JI. Certainly, the implementation of JI will face numerous challenges because so many operational aspects have to be addressed. Because of the space limitation, however, our discussion will focus on the following most important aspects.

---

[32] Facilitating immediate progress without jeopardizing a smooth evolution to a more mature, comprehensive regime requires careful attention to the implementation details. Although the joint implementation stage bears little resemblance to an actual emission permits market, it serves the very important purpose of ensuring a smooth evolution of the trading regime and providing opportunities for the various supporting administrative institutions to 'learn by doing'. See, for example, Tietenberg and Victor (1994) for further discussion of implementation issues for a global tradeable carbon permits regime.

2.5.6.1 Potential benefits of joint implementation

Greenhouse gases are the uniformly mixed pollutants, that is, one ton of a greenhouse gas emitted anywhere on earth has the same effect as one ton emitted somewhere else. Translated into the language of abatement strategies, this means that it does not matter whether greenhouse gas emissions are reduced in the United States or in China. What matters is whether we are able to reduce the emissions effectively on a global scale. This argument provides the environmental rationale for JI (cf. Loske and Oberthür, 1994; Pearce, 1995a). The environmental argument in favour of JI is further supported by the following legal basis for JI. As stated in Article 3.3 of the FCCC, 'efforts to address climate change may be carried out cooperatively by interested Parties'. Moreover, Article 4.2(a) states the developed country Parties and other Parties included in Annex I may implement[33] ... policies and measures jointly with other Parties and may assist other Parties in contributing to the achievement of the objective of the Convention (cf. Ramakrishna, 1994).[34] Furthermore, the first CoP to the Convention in Berlin in April 1995 endorsed a pilot phase of JI referred to as activities implemented jointly (AIJ) among Annex I Parties and, on a voluntary basis, with non-Annex I Parties (that is, developing countries). The pilot phase ends no later than the year 2000.

Until now, the most widely recognised benefit of JI is its potential to act to lower the costs of undertaking greenhouse gas abatement in the industrialized countries that are currently responsible for the majority of global CO_2 emissions and hence to reduce the competitive disadvantage and carbon leakage associated with purely unilateral policies in these countries (cf. Barrett, 1995; Jepma, 1995). Worldwide, this will achieve global abatement at a lower overall cost than would otherwise have been the case.

JI offers opportunities for the active involvement of the private sector, provided that financial or legal incentives to abate emissions are offered. This provides opportunities to attract additional funds from the private sector of the investor countries. Closely related to this, two points need to be made. First, the governments of Annex II countries[35] should not regard private JI projects as a substitute for current official development assistance (see Section 2.5.6.3 for a further discussion). Second, many countries like Denmark, the Netherlands have

---

[33] The Annex I countries refer to the OECD countries and Central and Eastern European countries. These countries are listed in Annex I to the FCCC and have committed themselves to emission targets.

[34] Articles 2.5 and 2.8 of the Montreal Protocol on Substances that Deplete the Ozone Layer can be viewed as a limited precursor to JI under the FCCC (cf. *International Legal Materials*, various issues).

[35] The Annex II countries refer to the OECD countries that are listed in Annex II to the FCCC and have committed themselves to emission targets.

proposed that emission reductions achieved during the pilot phase through JI projects should not be credited to current national commitments of investor countries under the FCCC. But crediting is an element unique to JI deals. Without crediting or other reward, JI projects are no different from traditional environmental aid and thus it is doubtful whether a large number of private JI projects will get off the ground (cf. Kuik *et al.*, 1994; Nordic Council of Ministers, 1995). Indeed, since inception of the pilot phase, a relatively small number (currently around 40) of AIJ projects has so far been officially reported to the FCCC Secretariat as being accepted, approved or endorsed by the governments of the host and investor countries. Moreover, the geographical distribution of these projects is quite uneven, with very few AIJ projects being established in Africa and Asia. Given the short time horizon of the AIJ pilot phase and the lack of a diversified base of the current AIJ projects, there would not be enough practical experience to provide an empirical basis for a decision on whether to move forward beyond the pilot phase.

As far as the developing countries are concerned, JI provides other positive environmental effects, since JI also helps to curb local pollution. The developing countries perspective on the benefits from a JI project is different from those of the investor countries, the latter of which regard abated global greenhouse gas emissions as the most important benefits from the project. Moreover, through participating in JI projects, the developing countries can get increased access to more advanced abatement technologies and additional funding, although the extent of participation would depend on the definition of 'incremental costs' of JI deals. This will make it possible for the developing countries to lower energy use and hence emissions while achieving the same rate of economic growth (that is, 'technological leapfrogging'). Furthermore, as discussed in Section 2.3, the developing countries are even more vulnerable to climate change, and a broad commitment to JI would also reduce the damage potential from climate change in the developing countries themselves, since after all it is not only the industrialized countries whose climate will change if greenhouse gas emissions are not reduced.

2.5.6.2 Forms of joint implementation

JI can be broadly defined as an attempt to reduce the global costs of meeting a particular GHG emission target. JI in a wider sense could cover more general cooperation between two or more countries on measures to abate GHG emissions, but this type of JI has up to now been addressed to only a limited extent in the international climate change debate.[36] Unless otherwise specified, the following discussion is therefore based on the current dominant definition of JI at the project level.

---

[36] This type of extended JI requires restrictive preconditions. See, for example, Nordic Council of Ministers (1995) for discussing the possibility of developing this type of JI within the Nordic countries.

There are three possibilities of introducing JI at the project level. First is a multilateral approach to JI through an entity such as Global Environment Facility (GEF) (cf. Merkus, 1992; Jones, 1994; Michaelowa, 1995). Specifically, countries wishing to invest in JI projects pool their resources to an independent fund, whereas other countries offering JI projects compete for the funding resources. During the duration of the selected JI project, each investor country receives the credit proportional to its share of the project portfolio. The major advantage of the multilateral approach is risk-sharing because project risks can be spread among all the investor countries. On the other hand, there are some disadvantages. For the sake of reducing administrative overheads, the approach results in a preference for large-scale projects. Moreover, because of the multilateral characteristics, the approach disregards the diversified preferences of each investor country. All this will reduce the diversity of JI projects. Furthermore, because the project selection and approval cannot avoid the dangers of international bureaucracy and abuse of power, the approach would have serious drawbacks for both the efficiency and equity of the JI market (Environmental Defense Fund, 1993; Dudek and Wiener, 1996).

Second, JI deals are through government agreements between two states (cf. Merkus, 1992; Jones, 1994; Michaelowa, 1995). In this form, JI contracts are concluded at a government level and executing JI projects can be commissioned to public entities at a national, regional or municipal level, or to private companies and organizations. In either case, state authorities must be informed of the progress projects are making before issuing a certificate of approval. Moreover, in order to reduce administrative costs, an institution could be established to act on behalf of the countries concerned. This would represent a form of clearing house (cf. Kuik *et al.*, 1994; Michaelowa, 1995; Nordic Council of Ministers, 1995). Such a clearing house would deal with the tasks, such as the identification of JI projects, spreading risk, reducing transaction costs, and the close follow-up of individual projects (cf. Nordic Council of Ministers, 1995). Clearly, this approach differs from the above-mentioned GEF approach because JI projects are not bundled together in a portfolio as in the GEF case (Michaelowa, 1995).

Third, JI deals can be carried out by the private sector (Jones, 1994; Nordic Council of Ministers, 1995). Private companies may become actively involved in JI projects, if financial or legal incentives for them to abate emissions are provided. To some extent, the incentive for their involvement could come from a 'first-mover advantage', which strengthens the international competitiveness (in world markets in the future) of such companies that take the lead in developing climate-benign technologies (Loske and Oberthür, 1994).[37] This may be particularly true for large companies. To a less extent, the involvement of the private sector is also because of a fear of new regulations at home and a desire for a

---

[37] See, for instance, Skea (1995) for general discussion of 'first-mover advantage' in the context of environmental technology.

positive environmental profile.[38] This type of JI provides opportunities to attract additional funds from the private sector. Given the limited amount of public funds available, this approach is considered particularly important in order to obtain the necessary investments in JI projects (cf. Metz, 1995). Moreover, the approach can bypass inefficient bureaucracies from which public projects often suffer, thus keeping transaction costs at a minimum. In addition to the private sector involvement, non-government organizations (NGOs) should be given the opportunity to participate in JI projects, but their activities should focus on capacity building, monitoring and certification rather than pursuing JI projects *per se*. However, NGOs are strongly opposed to the concept of JI, because they have accused the industrialized countries of using JI as a means of buying their way out of responsibility for climate problems and at the same time postponing the radical changes in their own consumption patterns and passing the responsibility on to the developing countries.[39] They will probably remain sceptical about JI unless they are convinced that clear criteria for JI have been established. Until then, it is therefore unlikely that the active involvement of NGOs in JI projects will emerge.

2.5.6.3 Criteria for joint implementation

Joint implementation is a climate policy instrument that may lead to comprehensive transfers of resources from rich to poor countries. From the beginning, however, the developing countries, with the support of western environmental NGOs, are strongly opposed to the concept of JI. If the potential for cost effectiveness and the transfer of resources is so large, why has it aroused so much opposition in the developing countries? This debate on JI has underlined the need to establish general criteria and conditions for how JI is to function.

According to the FCCC, the official criteria for JI will be laid down by the CoP. The type and size of the transaction costs of JI will depend on the criteria established by the CoP as well as the institutions and procedures designed to facilitate the development of JI projects.[40] We think that the essential criteria for JI should include the following, at least from the developing countries point of view.

---

[38] Nordic Council of Ministers (1995) thinks that this is the main reasons why private companies, mainly in the US, have carried out JI projects on a voluntary basis.

[39] The Climate Network Europe and the Greenpeace, for instance, hold a critical view on JI. See Climate Network Europe (1994) and Hare and Stevens (1995) for further discussion.

[40] The transaction costs of JI consist of search costs, negotiation costs, approval costs, monitoring costs, enforcement costs, and insurance costs. For a detailed discussion, see Dudek and Wiener (1996).

a) JI projects should be compatible with development priorities of the host countries. JI projects should bring about, in clear terms, real, measurable and long-term environmental benefits that would not have occurred in the absence of such projects. To this end, the prior acceptance, approval or endorsement by the national governments involved is deemed important, although this would add to approval costs. This is also in line with the Berlin Mandate, which states that 'all activities implemented jointly under the pilot phase require prior acceptance, approval or endorsement by the Governments of Parties participating in these activities'. Closely related to this, if one JI project is not compatible with development priorities of host countries, it is doubtful whether it can gain the host country acceptance, because only countries as a whole are the Parties to the Convention, and because JI projects are tied to agreements between governments. This is unique to JI projects compared with traditional development projects. Parikh (1995), for example, argues that reforestation projects should be rejected since they do not involve technology transfer and lead to potential conflicts with development priorities, especially land use. Moreover, it is not enough that JI projects be not harmful because harmless projects that are unrelated to development priorities divert limited resources away from priority activities and thus involve high opportunity costs for the host countries.

b) Funding for JI projects should be additional to current official development assistance of Annex II countries. In addition to emissions additionality, which requires that emissions should be reduced from what they would have been in the absence of the projects, the intent of financial additionality is that the funding for JI projects should not come from traditional development budgets packaged under a new name, because the developing countries generally fear that Annex II countries redefine existing development aid projects as JI projects and thus reduce their aid budgets accordingly, and because small developing countries particularly fear that Annex II countries tend their attention away from them towards those developing countries with great economies and greenhouse gas emissions.

In order to make sure that any resources for JI are additional, the Annex II countries should at least allocate a certain percentage of their gross national products (GNP) to official development assistance (ODA) (cf. Kuik *et al.*, 1994; Loske and Oberthür, 1994). If such an agreed on threshold cannot be established, it is very important to keep the funds used for JI projects clearly distinguishable from those of the existing ODA. Limiting the contributions of JI to domestic GHG emission reduction obligations in Annex II countries[41] as well as giving the Annex II only credit for part of the emission reduction achieved abroad may also help to reduce demand for JI projects and hence the incentives to shift the

---

[41] In order to lower transaction costs, such limits should be imposed on each investor rather than on the total national level. Moreover, they should differ per type of project.

funding from the existing ODA. Otherwise, the developing countries would probably remain sceptical about JI.

c) Priorities should be given to JI projects for limiting emissions over projects for enhancing carbon sinks. In the proposed criteria from Canada, the Netherlands, Norway, and the US, it has been stated that it should be possible for so-called sinks projects to become JI projects. Indeed, until now, projects for enhancing sinks through reforestation, afforestation or efficient forest use account for a large proportion of the existing projects that have been suggested to qualify in principle as officially recognizable JI projects (Environmental Defense Fund, 1993).[42] This is because projects of this type currently represent the least cost option. This may also be motivated by a concern to operate at a manageable level, with the goal of testing JI within the framework of the Climate Convention in order to gain concrete experience and to convince hitherto sceptical countries of the potential of JI projects.

By contrast, countries like Denmark have proposed that JI should not include sinks projects. The JI criteria from the Australian Pilot Phase JI Program also suggest that JI projects should reduce net greenhouse gas emissions. Their objections to including sinks projects appear mainly practical considerations, because there are the great uncertainties surrounding the true measures of carbon fixed, because there is the danger that countries may clear forests to have room for such JI projects later on, and because there are the high risks associated with such long-term sink-enhancing JI projects (cf. Bohm, 1994; Loske and Oberthür, 1994).[43] This may also be because sinks projects tend to merely postpone the problem of greenhouse gas emissions rather than solve it.

Faced with such sharp divergences in the proposed criteria, we think that priorities should be given to JI projects for limiting emissions, at the same time not excluding JI sinks projects. This stance can be explained briefly as follows.

While the investor countries regard abated global greenhouse gas emissions as the most important benefits from JI projects, a large number of host (developing) countries regard local environmental problems as their own environmental priorities (cf. Jones, 1994; Michaelowa, 1995). They are more concerned with local pollutants, such as SO_2, NOx and particulates from fossil fuel burning, because emissions of these pollutants cause serious health hazards and large environmental damage. Sinks projects have a favourable climate effect, but do not contribute to the reductions of these local pollutants and thus to solving local environmental problems.

---

[42] For a survey of the existing JI projects, see Pearce (1995b), Watt *et al.* (1995), and *Joint Implementation Quarterly* (various issues).

[43] Although the Netherlands does not reject JI sinks projects, it has suggested that common methods of evaluating this type of projects should be established first.

Moreover, the current emission stabilization target for industrialized countries are not sufficient to achieve the Climate Convention's ultimate objective of stabilizing GHG concentrations in the atmosphere at a level that would prevent dangerous anthropogenic interference with the climate system. Given the fact that developing countries are expected to experience emission increases in the coming decades and are not expected to make any new commitments going beyond the currently general ones under the FCCC, which may hinder their economic growth and development, the achievement of the ultimate objective will rely on strengthened obligations for industrialized countries to limit their own emissions or on JI projects for limiting emissions. Since the former is considered too costly and not cost-effective by industrialized countries, which is the reason for JI, main reliance should thus be placed on JI projects for limiting emissions in order to achieve the Climate Convention's ultimate goal.

These arguments by no means exclude JI sinks projects. Take deforestation as an example. All forests store carbon, but deforestation will release carbon dioxide into the atmosphere that will contribute to the accelerated greenhouse effect.[44] Brown and Pearce (1994) and Pearce (1996) have shown that the carbon storage value of forests are several times the domestic value. Thus, the sensitive, biologically diverse and rich forests of the developing countries could become a source of revenue not for timbering and clearing, but for preservation and enhancement. This, combined with global concern about tropical deforestation, suggests that avoiding deforestation through measures such as JI could become a potentially important means of reducing the greenhouse effect. Moreover, since tropical forests are generally located in tropical (developing) countries and since deforestation is mainly in those countries, opportunities for sink enhancement are generally largest in those countries, in which some industrialized countries, if not all, want to implement JI projects. Furthermore, avoiding deforestation is also in line with the national priorities of some of those countries. Thus, from a perspective of those countries, sinks enhancement for avoiding deforestation should not be excluded.

d) Guidelines should be established for the reporting of the performance of JI projects with respect to methodologies for calculating project baselines and actual emissions and for monitoring, verification and audit. The success of JI will critically depend on the ease with which JI projects can be arranged between interested parties. Standardizing the reporting procedures and requirements for JI projects would lower transaction costs and thus help to foster the development of JI projects. By placing emphasis on the documentation of all sources, methods, emission factors, and assumptions, it would also make it possible for an independent third party to validate the emissions estimates and project effects.

---

[44] The release rate of carbon dioxide differs, depending on the method of clearance and subsequent land use. For a further discussion, see Brown and Pearce (1994) and Pearce (1996).

2.5.6.4 The commitments of Annex I countries

The extent to which non-Annex I countries would work together with Annex I countries in implementing JI projects would depend on the Annex I commitments to be made at the upcoming third CoP to the FCCC scheduled to be held in Kyoto in December 1997. We think that such commitments should include the following:

First, Annex I countries should strengthen their existing commitments under the FCCC with respect to GHG emission targets and timetables, and transfers of financial resources, technology and expertise. Although much progress has been made since the 1992 Earth Summit in understanding the science of climate change, progress in the implementation of the FCCC has not been up to expectations. Indeed, most of Annex I countries continue to increase their emissions along an upward trajectory, which will result in the failing to meet their current commitments to returning their GHG emissions to their 1990 levels by the year 2000. Moreover, developing countries complain that Annex I countries have not lived up to the promise they made in Rio de Janeiro to help non-Annex I countries be greener. They continue to insist that Annex I countries must first meet their agreed commitments before non-Annex I countries will consider taking on additional commitments. The European Union (EU) broke the ice by offering a negotiation position of a 15% cut in emissions of a basket of three gases - CO_2, CH_4 and N_2O - below 1990 levels by 2010. The proposed target is for the EU as a whole, with targets for individual member states ranging from plus 40% for Portugal to minus 30% for Luxembourg. The EU proposal is the first formal one from Annex I countries, which contains a concrete target for emissions reductions. Although the proposal just is the EU negotiation position, not a commitment the EU will undertake on its own, it is seen by advocates of early action as a very crucial step in the right direction.

By permitting a 30-40% increase in emissions to Greece and Portugal, the EU proposal for internal community burden sharing accepts that poorer countries should be treated more leniently, although it has been considered inconsistent with the EU opposition to differentiated emissions targets among Annex I countries. But if Greece and Portugal can have this sort of rise, what leeway should be allowed for the really poor, that is, non-Annex I countries. Moreover, given the fact that many EU countries are still on an upward trajectory of GHG emissions since 1990, the proposal has raised the question whether the proposed emission reductions within the suggested time-frame are realistic. Besides, the EU and the US have been bickering. The US, with the backing of Australia, Canada and Japan, has been critical of the EU insistence on mandatory policies and measures as well as short-term targets. Although the US appears not to reject proposals for setting legally binding targets for emission reductions, the US is unlikely to agree to any targets unless it knows what flexibility it could have. Here flexibility refers to the following: (1) would the carbon permits be issued as an emissions budget over a period? (2) can early-achievement be banked for future use and can under-achievement in the current period be fulfilled by the borrowed permits from a subsequent period? and (3) can emission reductions be achieved 'offshore' through emissions trading or joint implementation? Clearly, the ongoing tension over the responsibilities of different parties to the FCCC suggests that if there would be any concrete commitments and emission targets specified at the upcoming third CoP, they would only be the result of negotiation among the parties themselves.

Second, Annex I countries should provide adequate domestic incentives to encourage their private sector participation in JI projects. JI can only be successful if there is the active involvement of the private sector in project financing.

2.5.6.5 Joint implementation and the baselines

Article 4.3 of the FCCC states that the developed country Parties shall provide financial resources, including the transfer of technology, needed by the developing country Parties to meet the agreed full incremental costs of implementing measures to control greenhouse gas emissions. The term 'incremental costs' makes it clear that the investor countries only pay the incremental cost incurred by the host countries in implementing JI deals. Here, incremental costs, which have been frequently used in discussing the cost of GHG abatement and will be defined by the CoP, are incremental to what the host countries otherwise would have done. This brings up the issue of the baseline.

By definition, the baseline refers to the path of GHG emissions without any JI project. The baseline is deemed necessary in order to measure emission reductions resulting from JI projects and ensure correct crediting between the parties concerned. This is because, by establishing the baseline, we reduce the danger of the so-called double counting, where both investor and host parties claim the right to deductions on the basis of the same reduction volume (cf. Merkus, 1992; Michaelowa, 1995; Nordic Council of Ministers, 1995). Moreover, by establishing the baseline, we reduce the free-rider effects. Otherwise, a JI deal might sanction a reduction that would have taken place anyway (Loske and Oberthür, 1994; Pearce, 1995a). Furthermore, establishing the baseline at the highest possible aggregated level, be it the national or even the international level, would reduce the leakage effects which occur when reduced GHG emissions in one place are counteracted by increased emissions elsewhere in the same host country or even in other countries as a direct or indirect effect of the JI project itself (Climate Network Europe, 1994; Watt *et al.*, 1995).[45] From this, it therefore

---

[45] Direct effect of a JI project means that, for example, the coal that is saved by a JI project aimed at improving the efficiency of a power station may be used for another power station in the same host country, while indirect effects refer to those effects that can arise as a result of the changes in relative prices and behaviour via a JI project.

follows that the suggestion that the baselines are not needed for JI deals under the FCCC is sceptical.[46]

For Annex I countries, the Climate Convention commits them to cut down emissions of CO_2 and other greenhouse gases to their 1990 levels by the year 2000. Since the baselines of Annex I countries are related to their historical (1990) emission levels, there is little uncertainty about their future levels. However, for the developing countries with no abatement commitments under the FCCC, establishing their future national baselines is not a simple matter. First of all, this is because the baseline can never be actually observed. Put another way, it is impossible to observe what would have happened if JI deals had not been implemented.

Second, the definition of the baseline itself is not without conceptual problems (cf. Merkus, 1992; Climate Network Europe, 1994; Michaelowa, 1995; Nordic Council of Ministers, 1995). Given the fact that climate policy in the developing (host) countries is not so much about absolute emission reduction but about slowing the rate of growth of future GHG emissions, the host countries tend to 'inflate' their baseline scenarios and regard any effort to reduce the growth of emissions as incremental. By contrast, the industrialized countries argue that the baselines in the host countries should be adjusted by eliminating projects that would have been carried out anyway by the host countries themselves and by subtracting emissions induced by energy subsidies and other economic distortions. As a result, they come to a much lower emissions baseline (cf. Loske and Oberthür, 1994). Clearly, there exists much controversy about the extent to which policy distortions and abatement projects should be included in the baseline. As noted earlier, given the fact that the investor countries only pay the incremental costs of JI projects in the host countries, it is therefore essential to come to a consensus on the baselines.

Third, there are great uncertainties surrounding the baseline. Fritsche (1994) shows that the variation in the baseline that is established for CO_2 emissions in the European Union by different approaches is in the order of at least 10% of the national emissions. The variation in the baseline tends to be even greater for CO_2 emissions in the developing countries and for emissions of other greenhouse gases but CO_2. This underlines the need to establish a common methodology on the baseline.

Fourth, from the point of view of strategic behaviour, the developing countries may be even unwilling to establish their baselines because doing so may convey the impression that they would bind themselves to these aggregate emission paths.

At present, no developing countries have established national emission targets. Moreover, it is unlikely that these countries will adopt binding targets in the near future. This underlines the great uncertainty of obtaining an accurate evaluation of emissions for JI deals. Then the question arising from this is whether JI

---

[46] Jones (1994), for example, suggests that no attempt should be made to determine the baselines.

projects should be limited to countries that have national emission targets. The argument in favour of JI projects between countries that both have emission targets is that the requirements for measurement and control of JI projects are in any case reduced. Moreover, there is a great certainty that such projects will contribute to the reduction in global emission. On the other hand, the argument for not limiting JI only to countries with national emission targets is that the developing countries, where the potential for cheap emission abatement JI investments is far greater than in Annex I countries, will be not excluded. In this case, the baselines at project level at least have to be developed in order to suffice for JI arrangements.

2.5.6.6 The verification of GHG emission reductions of JI projects

Both the investor country and the host country have incentives to inflate the effect of JI projects. The investor countries may be tempted to inflate the volume of emission reduction from JI projects in order to receive greater credit than the JI projects merit, while the host countries may be tempted to inflate the potential for emission reduction from current JI projects in order to attract future JI projects given the fact that the investor countries attempt to get as much as possible out of their investment. Moreover, there are great uncertainties associated with the measurement of the actual emission reduction of a given JI project itself (cf. Climate Network Europe, 1994). All this, combined with the complexities in establishing the baseline, underlines the need for the verification of the GHG emission reduction in order to ensure correct crediting.

Since both the investor party and the host party have an incentive for exaggerating the emission reduction, it is particularly desirable that the verification is carried out by an objective third party that should be agreed to by both partners. The verification is only responsible for deciding whether to accept the calculated GHG emission reduction or not. It is not the purpose of the verification to evaluate the acceptability of the JI project. It is up to the host and investor countries to decide what their definition of a JI project is (cf. Watt *et al.*, 1995).

For a given JI project, the extent of verification requirements depends on its characteristics and duration. *Ceteris paribus*, the more intensive the verification, the higher the transaction costs associated with measurement and inspection of the JI project. In order to reduce the costs, it is therefore desirable to establish the standardized verification method for each category of JI projects and to make use of the existing, suitable institutional apparatus as much as possible (cf. Nordic Council of Ministers, 1995).

Since the pilot JI projects are just being launched, verifying the GHG emission reduction of JI projects is still at the initial stage. Moreover, the great differences exist in the national conditions among the host countries. Thus, it is not surprising that a wide range of possible institutions from fairly decentralized to quite centralized have been suggested (cf. Watt *et al.*, 1995). In examining the roles for potential institutions in China, for example, Zhou and Li (1995) suggest that the Energy Research Institute of the State Planning Commission could function as a third party to evaluate energy-related JI projects, and that the Chinese Academy of Environmental Sciences of the National Environmental Protection Agency could inspect those JI projects aimed at environmental control. Moreover, on the basis of sovereignty considerations, they do not believe that the Chinese government would react positively to international verification, although the limited involvement of some sort of United Nations' related team would be acceptable. Clearly, verification of this type is of the centralized structure. It would be preferred to a decentralized project-related verification because the former is carried out by means of the standardized procedure. However, the central solution at the national level cannot avoid the dangers of bureaucracy and abuse of power and is costly in comparison with a decentralized solution, because every JI project has to be evaluated by a single authority. On the other hand, it allows continuous improvement of the verification method by learning the lessons from failed JI projects (cf. Michaelowa, 1995).

In verifying the GHG emission reduction, it is conceivable that the event of disagreement about the results of a verification could arise. Thus, it is essential for the CoP to establish a dispute settlement procedure that could be based on the FCCC multi-lateral consultative mechanism, or the independent panel model currently being used in the World Bank, or other mechanism. Whatever the procedure that is eventually established, it should be made available for all disagreements about the verification results brought by any host and investor parties. Once the dispute is settled, sanctions can therefore be fairly imposed on the parties for breaches of contract.

2.6   Concluding remarks

Climate change is expected to have a variety of consequences. The great bulk of the scientific work has so far focused on the case of a doubling of concentration of carbon dioxide equivalent trace gases from its pre-industrial level. The studies discussed in this chapter show that agriculture loss and loss from a rise in the sea level form the main sources of damage due to a doubling of CO_2 concentration, although the effects on the supply of water, health, ecosystems and the energy sector are also important. With respect to the regional effects, there is much less agreement on the exact magnitude and timing of regional damages from climate change, although there is considerable agreement that less developed regions will suffer more than their developed counterparts, with China being the region hardest hit.

Although the great uncertainties by no means justify the 'do nothing' stances, they indeed make it difficult to make the choice of response strategies. This can be reflected by the following divergences:

First, 'do nothing' versus 'do something'. The 'do nothing' stances argue that we should wait until considerable consensus in scientific research emerges. By contrast, the 'do something' stances argue that we can probably not afford to wait until all of the uncertainties have been removed.

Second, 'no-regrets' policies versus 'regrets' policies. Advocates of 'no-regrets' policies argue that strong action will lead to a 'no-regrets' outcome, which is independent of the ultimate seriousness of global climate change. By contrast, advocates of 'regrets' policies argue that emission abatement is very costly and that the expected costs of such action would not be justified unless there were going to be adverse impacts of the greenhouse effect. Moreover, even within the group of proponents of 'regrets' policies, divergences occur in choosing such policy instruments as carbon taxes or tradeable carbon permits.

Third, limitation strategies versus adaptation strategies. The former aim to limit, stop or even reduce the growth of GHG concentrations in the atmosphere, whereas the latter are to adjust the environment or the ways it is used to reduce the consequences of a changing climate.

Such sharp divergences in the suggested policy strategies are hardly expected to emerge in the near future unless there is considerable consensus in scientific research. Faced with such sharp divergences, the Manne and Richels 'hedging' strategy will tend to be favoured. Such a strategy avoids extreme and rigid mandates and pursues a number of low-cost options for preparing the ground for more severe measures later if needed. The strategy also makes it easy to adapt to climate change by cutting GHG emissions, since adaptation requires that any impact, such as a rise in sea level, should occur sufficiently slowly to allow orderly adjustment to take place.

If the strategy were adopted, it should be pursued in the most cost-effective manner. Section 2.5 discusses how this goal can be achieved in the context of CO_2 emissions. The following conclusions emerge from that discussion.

First, if the goal is to reduce CO_2 emissions, energy taxes (if introduced) will lead to poor target achievement or else to unnecessarily high costs as compared with carbon taxes. In the case of general taxation on energy, the reductions in CO_2 emissions will be mainly achieved by price-induced energy conservation. By contrast, carbon taxes reduce CO_2 emissions through the effects of price mechanism on energy consumption and fuel choice. Given the cost-effectiveness of a carbon tax, the computable general equilibrium model of the Chinese economy, which will be described in Chapter 5, incorporates it as a policy instrument to reduce CO_2 emissions in China.

Second, for the effectiveness of action, carbon taxes should increase over time and be imposed globally in order to reflect the rising cost of damage resulting from the accumulation of CO_2 concentration in the atmosphere, to give the markets the signal that CO_2 emissions will eventually be heavily taxed, and to prevent carbon leakage that would otherwise take place in regions or countries without such taxes. But if this is not the case, special attention should be given to the treatment of the carbon tax revenues, the impacts on the distribution of income, and to the effects on international competitiveness when designing a unilateral carbon tax.

Third, the allocation of emission permits would, in practice, have to be based on relatively straightforward rules. If an allocation rule is likely to induce relatively large participation, account should be taken of historical CO_2

emissions, GNP (or GDP) and population together, and the emission entitlements to each participating country should be adjusted over time in order to reduce the relative benefits and relative excess costs for each country.

Fourth, the actual achievements in the reduction of CO_2 emissions by a proposed carbon tax remain uncertain, while under a regime of tradeable carbon permits there will be certainty about the magnitude of emission reductions but great uncertainty about the cost of such reductions. If there was a threshold effect of climate change, tradeable permits would be preferred to carbon taxes. Given the current lack of knowledge about the magnitude, timing and regional effects of climate change, however, carbon taxes appear to be the superior and more flexible instrument that avoids potentially large and unexpected costs.

Finally, although the joint implementation stage bears little resemblance to an actual emission permits market, it is worthwhile putting joint implementation into practice. Testing JI at a manageable level within the framework of the Climate Convention will gain concrete experience, convince hitherto sceptical countries of the potential of JI projects, and eventually serve the very important purpose of ensuring a smooth evolution of the trading regime and providing opportunities for the various supporting administrative institutions to 'learn by doing'. However, given the breadth of the subject of JI and its close linkage with national sovereignty, global political agenda, national development priorities, access to advanced technologies and development assistance, and the ongoing tension over the responsibilities of different parties to the FCCC, it is expected that a wide and successful implementation of JI will be conditional upon the consensus on a variety of operational issues such as transaction costs, the form of JI, criteria for JI, the establishment of baselines against which the effects of JI projects can be measured, and the verification of emission reductions of JI projects.

3 Analysis of the Chinese Energy System: Implications for Future CO_2 Emissions[1]

3.1 Introduction

At present, as the world's most populous country and largest coal producer and consumer, China alone contributes 11% of global CO_2 emissions (Manne and Richels, 1991a). This means that China ranks second in global CO_2 emissions if the Soviet emissions are distributed over the new independent republics. Under a business-as-usual scenario, China's contribution to global CO_2 emissions is estimated to rise to 17% in 2050 and to 28% in 2100 (Manne, 1992). Thus, advocates of controlling CO_2 emissions call for substantial efforts in China. However, the Chinese authorities have argued that China cannot be expected to make a significant contribution to solving the carbon emission problem unless China receives a very large amount of international aid for this purpose. This contrasts sharply with the wishes of proponents of controlling CO_2 emissions. To explain this gap a better understanding of the Chinese energy system is needed. This chapter is devoted to serving such a purpose by examining some aspects of the Chinese energy system and, at the same time, sheds light on the implications for China's future CO_2 emissions.

The structure of this chapter is as follows. Section 3.2 discusses China's energy resources and their development. In Section 3.3, the Chinese energy consumption patterns are characterized. Section 3.4 deals with electricity generation. Section 3.5 reviews China's energy conservation in an international perspective. The analysis of historical CO_2 emissions in China is presented in Section 3.6, while environmental challenges for the Chinese energy system are described in Section 3.7. This chapter ends with some observations from global studies about China's potential importance as a source of CO_2 emissions.

3.2 Energy resources and their development

In this section, I will discuss the reserves of coal, oil, natural gas, hydropower, uranium and renewables and their development. Figure 3.1 can be used for locating Chinese provinces referred to in this book.

---

[1] This chapter is to a large extent based on three papers in *International Journal of Environment and Pollution* (Zhang, 1994b), *Energy Policy* (Zhang, 1995b), and in *Intereconomics* (Zhang, 1996a).

Source: The Economist, March 18th-24th, 1995.

Figure 3.1   *Location of China's provinces*

3.2.1 Coal

China has abundant coal resources. At the end of 1993, its proven coal reserves
of 1001.9 billion tons were among the largest in the world, representing alone
about 50% of the world total (State Economic and Trade Commission, 1996;
World Resources Institute, 1992). Of the coal reserves, bituminous coal accounts
for 75%, anthracite for 12%, and lignite for 13% (State Economic and Trade
Commission, 1996).

In terms of 'proved amount in place' as defined by the World Energy Conference, however, China's position may look somewhat less dominant. As shown in
Table 3.1, only a small proportion of China's proven coal reserves, namely,
114.5 billion tons by the end of 1992, was classified as proven recoverable
reserves. This amounts to only half of the reserve estimates for either the US
(249.2 billion tons) or the former USSR (241.0 billion tons) (British Petroleum,
1993). Moreover, the geographical distribution of China's coal reserves is quite
uneven, with 72.5% of proven coal reserves concentrated in the northern part of
the country (see also Section 3.3).

3.2.2 Oil

According to the national oil and gas resources assessment completed in 1993, the total oil reserves in China were 94.0 billion tons (State Economic and Trade Commission, 1996). Its proven recoverable reserves of oil were estimated at 3.2 billion tons, or 5.6% of its proven recoverable reserves of coal (British Petroleum, 1993). At the 1992 level of production, this yields a reserves-to-production ratio of 22.2 years.

Table 3.1   *Reserves and utilization rates of fossil fuels*

Resources	Reserves	Proven recoverable reserves[1]	R/P ratio[2] (years)		Per capita proven recoverable reserves[3]	
	China	China	China	World	China	World
Coal (10^9 t)	1001.9	114.5 (11.0%)	103.2	232.0	96.4	189.7
Oil (10^9 t)	94.0	3.2 (2.4%)	22.2	43.1	2.7	24.9
Natural gas (10^{12} m^3)	38.0	1.4 (1.0%)	92.6	64.8	1178	25242

[1] Figures in parentheses denote percentage of world total.
[2] R/P ratio stands for the lifetime of proven recoverable reserves at 1992 rates of production.
[3] Measured in tons for coal and oil, and in cubic metres for natural gas.

Sources: British Petroleum (1993); State Economic and Trade Commission (1996); United Nations (1993).

Proven oil reserves are mostly concentrated in the north and northeast, which are the traditional oil producing regions. Since these traditional fields have all reached the mature stage, the discovery of additional oil reserves is now deemed vital. To this end, increasing investment in oil exploration is needed in order to cope with the rising costs of raw materials and equipments required for oil exploration (see Table 3.2). This also holds for oil production (see Table 3.2).

3.2.3 Natural gas

Natural gas is of paramount importance because its carbon emissions are lower than those from coal and oil. Theoretically, the future of natural gas is very promising in China. China National Petroleum Corp. estimated the total reserves at 38 trillion cubic metres (State Economic and Trade Commission, 1996). Because of the lack of exploration work, however, the proven recoverable reserves were estimated at only 1.4 trillion cubic metres in 1992, or 1.0% of the world total (British Petroleum, 1993). This represents a reserves-to-production ratio of 92.6 years; when compared with 16 years of the OECD, which accounted for 9.8% of the world total (British Petroleum, 1993), China's higher ratio is a reflection of its current lower production level. Moreover, these Chinese gas reserves are nearly all concentrated in Sichuan province.

Table 3.2   *Exploration investment and production costs of oil*

	1980	1985	1988	1989	1990
Exploration investment (yuan/t)[a,b]		491.0[c]	806.5	937.6	1225.4
Production costs of oil (yuan/t)[a]	43.6	61.2	97.6	144.1	177.6

[a] Measured at constant prices.
[b] Required to add an incremental ton of oil to the proven reserves.
[c] Average figure over the period 1981-85.

Source: Chen *et al.* (1993).

Historically, natural gas has been considered a low priority in China's energy sector. Over recent decades, the share of natural gas in the total investment in oil and gas has accounted for about 10%, much smaller than 22% in the former USSR between the period 1981-85 (Wang, 1992). This has had the following consequences:

1. low level of exploration for natural gas. By the end of 1992, the proven reserves of natural gas accounted for only 4.2% of its total estimated reserves, much lower than the world average of 54.8% (Wang, 1992; British Petroleum, 1993);
2. great disparity between gas and oil. In 1992, the world equivalent production ratio of natural gas to oil was 0.58:1 (British Petroleum, 1993). Gas production in some countries, such as the former USSR, USA and Canada, even surpasses their oil production. But the corresponding figure for China was only 0.10:1, about one-sixth of the world average (British Petroleum, 1993). In terms of proven recoverable reserves, the equivalent ratio of natural gas to oil in 1992 was 0.40:1 in China, about half the world average (0.93:1) (British Petroleum, 1993);
3. slow growth of production and proven reserves of natural gas. Over the period 1981-90, the average annual growth rate of world gas production was 3.2%, whereas the corresponding figure for China was only 1.5% (British Petroleum, 1990 and 1993). China's proven reserves of natural gas have also been upgraded more slowly than the world total.

Natural gas was previously regarded primarily as feedstock for the petrochemicals industry, but now it is being recognized as an alternative source of domestic energy supply. This has brought about increased exploration activity aimed at boosting the production to 30 billion cubic metres by 2000 from 17.6 billion cubic metres in 1995 (State Economic and Trade Commission, 1996).

3.2.4 Hydropower

There are abundant hydropower resources in China. Its economically exploitable capacity is estimated at as much as 378 gigawatts (GW) - the largest in the world - that would generate 1920 terawatt-hours (TWh), which is three times the 1990 level of electricity production (Zhang, 1991). However, nearly 70% of the total exploitable potential is located in the southwest, remote from load centres, thus resulting in unfavourable exploitation conditions.

Moreover, China has an exploitable capacity of small hydropower projects of 76 GW, corresponding to 250 TWh of annual electricity production (Chen, 1991). The development of small hydropower projects is an appropriate solution to meet the power needs for industrial and domestic use in rural areas.

Hydropower is a carbon-free energy resource. With the world's largest exploitable potential, China should make hydropower its energy resource of choice in consideration of CO$_2$ emissions. By the end of 1994, however, only 13% of the exploitable potential had been developed. Meanwhile, the total capacity installed of small hydropower plants amounted to 15.65 GW, only 20.6% of the corresponding exploitable potential. This means that there is great potential for hydropower development in China (see Section 3.4).

3.2.5 Uranium and nuclear power

Exploration for uranium in China began in the mid-1950s. Its reported uranium resources are mainly found in the Shanxi and Xinjiang provinces. Economically exploitable reserves of uranium are estimated to be a minimum of 100,000 tons, with total known reserves in excess of 800,000 tons (Owen and Neal, 1989). The latter is approximately equivalent to Australia's total known reserves, which are the largest in the Western world (Owen and Neal, 1989). Thus, the domestic uranium supply will not be a limiting factor in China's short-term nuclear power development.

At the time of writing, two nuclear power stations have been commissioned based on the most matured commercial pressurized-water reactors (PWRs), specifically, Qinshan Nuclear Power Station in Zhejiang province and Daya Bay Nuclear Station in Guangdong province. The former, with a capacity of 300 megawatts (MW), is Chinese-designed and was successfully put into parallel operation with the East China Power Network in December 1991. The latter has two French-designed 900 MW units, with the first unit commissioned in February 1994 and the second in May 1994. This marked the start of the development of nuclear power in China.

With the exception of hydropower, nuclear power has so far provided the only proven method for enormous potential for a large-scale generation of electricity without a parallel production of CO$_2$ emissions. But, in many parts of the world the future of nuclear power is clouded by the risks of accidents, the hazards of radioactive waste disposal and public acceptability. This leaves the future of nuclear power uncertain. In China, however, no significant objections to nuclear power development exist on either of the above-mentioned grounds (Zhou, 1992). Indeed, the economically developed coastal areas and other energy-deficient areas are eager for nuclear power stations being built in their territories.

A rapid development of nuclear power in China is expected to take place in the first half of the next century in order to ease the increasing energy shortages, with total capacity installed estimated at hundreds of gigawatts (cf. Lu, 1986; He and Lu, 1992). For such a large-scale development, the demand for uranium will be of the order of several million tons. This raises serious concern about availability of domestic uranium resources, which, as discussed earlier, is almost an order-of-magnitude lower than that required. From both the economic and political points of view, such a concern would indicate that the long-term development of nuclear reactors should shift from PWRs to advanced PWRs and fast breeder reactors. Compared with PWRs, which can make use of only about 1-2% of the energy content of uranium, breeder reactors can make use of up to 60-70% of natural uranium by efficiently converting uranium-238 into plutonium-239, which, like uranium-235, is a good reactor fuel (CERS, 1986; Williams, 1992). With breeder reactor technology, domestic uranium resources would be adequate to support a large-scale commitment to nuclear power, which would yield multiple benefits, including a reduction in CO_2 emissions.

3.2.6 Renewables

Renewables encompass solar, wind, geothermal and tidal energies. Over the past few years, their total supply has amounted to only 0.43 million tons of coal equivalent (Mtce) per year, accounting for 0.04% of the total national commercial energy supply (Wu and Wei, 1992). As described below, China is abundant in renewables. This abundance, combined with energy shortages in China as a whole and in the rural areas in particular, suggests that much attention should be paid to the development of renewables to supplement conventional energy resources in the long-term energy plan.

3.2.6.1 Solar energy
China is abundant in solar energy resources. About two-thirds of its territory receives solar radiation of above 140 thousand calories per square centimeter (kcal/cm^2) for over 2000 hours per year (CERS, 1986). In large parts of Qinghai and Tibet, the annual amount of solar radiation even reaches 220 kcal/cm^2 (CERS, 1986).

Development of a wide variety of technologies for solar energy utilization has begun in China. Among equipment using solar energy are solar water heaters, solar cookers, solar greenhouses, solar desiccators, passive solar houses, and photovoltaic devices. By the end of 1990, 1.8 million square meters (m^2) of solar heat collectors had been installed; there were 0.12 million sets of solar cookers in use with each, if maintained well, saving 500 to 1000 kilograms of fuelwood annually; 110 sets of solar desiccators had been used with a total aperture area of 11,000 m^2; about 800 sets of passive solar houses had been built with a total floor area of 0.3 million m^2; and 1.5 MW of photovoltaic cells had been used on land primarily as a source of electricity for railway signals, remote communications equipment, and livestock fences (Chen, 1991; State Planning Commission, 1992; Wu and Wei, 1992).

3.2.6.2 Wind energy
China's southeastern coast and some nearby islands boast abundant wind resources with a density of wind energy greater than 200 watts per square metre. In these areas, wind speeds from 3 to 20 metres per second (m/s) are available for 7000 to 8000 hours per year, with an annual average of 6-9 m/s for 4000 hours (Wu and Wei, 1992; State Planning Commission, 1993). This is suitable for the development of medium and large wind-power generators, which can then be integrated into local power grids or operated together with diesel generators. In addition, there are abundant wind resources in remote and thinly populated Inner Mongolia, Xinjiang and Qinghai, where a vast area is beyond the reach of power grids, and the application of small wind-power generators designed for household lighting and television is favourable. For the country as a whole, exploitable wind energy is estimated at 253 GW (State Economic and Trade Commission, 1996).

By the end of 1990, 110,000 small wind-driven generators and 6 wind farms were in operation, with a corresponding total capacity installed of 12 MW and 4.5 MW respectively (State Planning Commission, 1992). There were also 1600 wind pumps in operation, with a total capacity installed of 2.1 MW (State Planning Commission, 1992).

3.2.6.3 Geothermal energy
Geothermal energy resources are spread throughout the country. Until the present, over 3000 outcrops of hot springs have been located, 90 of which are high-temperature (above 150°C) geothermal sites. These high-temperature geothermal resources are mainly concentrated in southern Tibet and western Yunnan. According to preliminary estimates, the total capacity of geothermal energy reaches 3200 GW, 3.5 GW of which can be used for electricity generation (Yang, 1992).

By the end of 1990, the total capacity installed of geothermal power generation had been expanded to 21 MW, most of which had been installed in Yanbajing Geothermal Power Station in Tibet (Chen, 1991; Wu and Wei, 1992); the annual direct use of low-medium temperature (below 100°C) geothermal resources amounted to 7.2 petajoules (PJ), much more than the geothermal power generation (State Planning Commission, 1992).

3.2.6.4 Tidal energy
China's tidal power reserves are concentrated in Fujian and Zhejiang provinces along the southeast coastline. The exploitable tidal power potential is estimated at 21.7 GW (State Economic and Trade Commission, 1996). At present, there are eight tidal power stations in operation, with a total capacity installed of 6.12 MW

(Yang, 1992). Among the largest is Jiangsha Tidal Power Station, with a capacity of 3.2 MW and an annual power production of 6 gigawatt-hours (GWh) (Chen, 1991).

A feasibility study has shown that the investment costs per kW for a medium-sized tidal scheme would be two to four times higher than that of small hydropower installations (Chen, 1991). This implies that, at this stage, no medium-sized tidal power project could be initiated without subsidies from the government.

3.3   Energy consumption patterns

Chinese energy consumption patterns can be characterized as follows:

Main reliance on domestic energy resources

China is self-sufficient in energy. Its entire economy is based on domestic energy resources. Even in the 1950s when oil was imported, 97% of energy supply was still from domestic sources. China's energy balance was also unaffected by the first rise in world oil prices. In 1990, Chinese energy imports and exports were 13.1 Mtce and 58.8 Mtce, with the former accounting for 1.3% of the total national commercial energy consumption and the latter for 5.7% of the total national commercial energy production respectively (State Statistical Bureau, 1992b).

Coal-dominant structure of energy consumption

China is one of the few countries in the world that relies on coal as its major source of energy. Over the past few years, coal has accounted for more than 75% of the primary energy consumption (State Statistical Bureau, 1992b). This share has remained stable after having increased from 70% in 1976 (see Figure 3.2). In 1990, coal provided 73% of the fuel and power required for industry, and 80% of commercial energy for household use. To a large extent, heavy reliance on coal is due to the domestic coal endowments. Restrictions on energy trade also play a role in this massive use of coal.[2] This high proportion of coal consumption leads to low efficiency of energy use, produces a large amount of

---

[2] International trade in energy, particularly in oil and natural gas, is controlled by the central government and national self-sufficiency in energy until now has been the policy, which limits the use of oil and gas. Although China's international trade system has already undergone a massive change in recent years, energy trade reform has lagged behind. Nevertheless, as part of China's drive to rejoin the World Trade Organization, the restrictions on international trade in energy are expected to be gradually removed.

CO$_2$ emissions (see Section 3.6), and gives rise to serious environmental pollution (see Section 3.7).

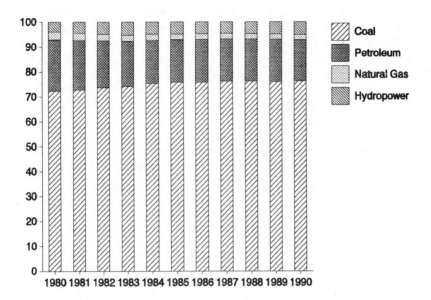

Source: Based on data from the State Statistical Bureau (1992b).

Figure 3.2   *Profile of energy consumption by fuel source, 1980-90*

Uneven geographical distribution of energy resources and economy

As shown in Table 3.3, 72.5% of existing coal reserves are concentrated in the northern part and 77.7% of exploitable hydropower potential is in the western part, while economically developed regions are located in the south and on the eastern coast. As a result, coal has to be transported over a long distance to the load centres. This in turn puts great pressure on the transportation system, especially on the severely congested railways. In 1989, 43% of the total freight shifted by railways was coal (Yang, 1991). This uneven geographical distribution also requires a major expansion of the transmission lines and power networks to realize 'sending electricity from the west to the east as well as from the north to the south' (cf. Ministry of Energy, 1992b).

Low per capita energy consumption but high energy use per GNP

The Chinese per capita commercial energy use in 1990 was 869.5 kilograms of coal equivalent (kgce), only about one-third of the world average, and its energy intensity measured as energy consumption per unit of GNP was 4.43 kgce/US$ (at the average exchange rate of 1990, 1 US$ = 4.78 Chinese yuan), among the highest in the world. The former is due to China's low level of development, and the latter reflects an unusually large share of energy-intensive industrial production in the Chinese economy, a large share of energy-intensive manufacturing in China's industry, a high proportion of coal consumption, and the undervaluation of China's GNP (see Section 3.5). This dual character of energy consumption in China has long been the crux of its energy problems.

Table 3.3 *Regional distribution of energy resources, economy and energy use in 1990 (%)*

	North China	Northeast China	East China	Central South	Southwest China	Northwest China
Energy resources of which:	32.3	5.9	9.6	8.5	23.7	20.0
Coal	43.2	5.8	11.4	6.2	9.9	23.5
Hydropower	1.2	2.0	3.6	15.5	67.8	9.9
Oil & Natural gas	10.0	47.8	18.4	8.0	4.7	11.1
GNP in 1990	13.5	11.7	33.2	25.3	10.6	5.7
Energy use in 1990 of which:	18.9	17.5	25.2	19.4	10.9	8.1
Coal	22.4	17.8	24.4	17.1	10.9	7.4
Electricity	16.5	15.3	29.0	21.6	9.3	8.3

Sources: State Statistical Bureau (1992a, 1992b); State Economic and Trade Commission (1996).

Heavy reliance on biomass energy by rural households

About three-quarters of the energy for domestic use of more than 800 million inhabitants in rural areas depends on biomass energy. In 1990, 215 million tons of firewood were burnt, 75 million tons more than that of the rational cutting (State Planning Commission, 1993). This overcutting has caused serious ecological damage to large areas of forest and vegetation.

Industry-dominant composition of final energy consumption

Since great differences exist in the economic structures between China and the industrialized countries, the compositions of their final energy consumption by sector are also quite different (see Table 3.4). In China, industry is the dominant energy-consuming sector, accounting for 67.1% of the total in 1990. With the shift towards a less energy-intensive society as the national economy grows, and the more energy-efficient industrial utilization, its share is expected to decline slightly in the future. The residential sector is at present the second largest user, consuming about 17% of the total final use in 1990, whereas the transport sector consumed only about 5% of the total in 1990. A similar picture applies to electricity consumption; industry consumed 78.2% of the electricity production in 1990, the residential sector 7.7%, and the transport sector 1.7% (State Statistical Bureau, 1992b).

Table 3.4   *Final energy consumption by sector (%)*

	China			USA	FRG	UK
	1980	1985	1990		1987	
Agriculture	6.0	5.5	5.1			
Industry	66.6	65.2	67.1	34.2	34.8	31.6
Construction	1.7	1.8	1.3			
Transport & Communication	5.0	5.0	4.7	35.0	23.6	27.9
Services	4.1	4.4	5.0			
Residential	16.6	18.1	16.8			
Others [1]	-	-	-	30.8	41.6	40.5
	100.0	100.0	100.0	100.0	100.0	100.0

[1]Including residential, commercial, public and agricultural sectors.

Sources: Based on data from the State Statistical Bureau (1990c, 1992b) and from the IEA (1989).

3.4   Electricity generation: achievements and remaining problems

As shown in Table 3.5, the electric power industry has been growing rapidly over the past decade in order to cope with the fast growing demand for electricity. During the Seventh Five-year Plan Period (1986-90), the total additions to generating capacity amounted to 50.84 GW, with average annual additions of 10 GW. This achievement made total national capacity installed reach 137.89 GW by the end of 1990, with thermal power accounting for 73.9% and hydropower for 26.1%. In addition, on 15 December 1991, the first domestically designed and constructed nuclear power unit with a capacity of 300 MW - Qinshan Nuclear Power Station - was successfully put into parallel operation with the East China Power Network (Ministry of Energy, 1992b). This marked the end of an era without nuclear power in China.

Along with a great number of units commissioned into operation, the total national electricity generation reached 621.3 TWh in 1990, more than double that of 1980. Power generated from thermal and hydropower plants accounted for 79.7% and 20.3% respectively, of the total electricity production. The share of coal-fired plants in total thermal generation rose from 73.6% in 1980 to 89.1% in 1990 as a result of government policy of replacing oil by coal in power generation in order to release much-needed oil into the economy. This share will continuously increase hereafter because oil-fired units will no longer be constructed generally and oil-fired units originally designed for burning oil will be re-converted to coal-fired ones (cf. Zhang, 1991).

Table 3.5   *Total capacity installed and electricity generation, 1980-90*

Year	Capacity installed			Electricity generation			
	Total (GW)	Hydro (%)	Thermal (%)	Total (TWh)	Hydro (%)	Coal (%)	Oil/gas (%)
1980	65.87	30.7	69.3	300.6	19.4	59.3	21.3
1981	69.13	31.7	68.3	309.3	21.2	58.7	20.1
1982	72.36	31.7	68.3	327.7	22.7	59.1	18.2
1983	76.44	31.6	68.4	351.4	24.6	59.1	16.3
1984	80.12	31.9	68.1	377.0	23.0	62.7	14.3
1985	87.05	30.3	69.7	410.7	22.5	64.6	12.9
1986	93.82	29.4	70.6	449.6	21.0	68.6	10.4
1987	102.90	29.3	70.7	497.3	20.2	70.4	9.4
1988	115.50	28.3	71.7	545.1	20.0	69.9	10.1
1989	126.64	27.0	73.0	584.7	20.3	70.4	9.3
1990	137.89	26.1	73.9	621.3	20.3	71.0	8.7

Source: Ministry of Energy (1991b).

Despite the impressive achievements in electricity generation, China is still a country with a low penetration of electricity in total final energy consumption (TFC). By 1990, the share of electricity in TFC was only 7.6%, considerably smaller than the average for the International Energy Agency (IEA) countries (16.6% in 1987) and also below many developing countries, for example, India's 15% and Brazil's 19% (State Statistical Bureau, 1992b; IEA, 1989; Hu, 1991). Such a small share is one consequence of a small proportion of coal used for electricity generation, which was only 25.8% of the total national coal consumption in China in 1990 compared with about 85% in the United States (Zhang, 1991; State Statistical Bureau, 1992b). This implies that coal utilization in China is quite inefficient because the majority of coal is consumed by direct combustion. Moreover, given that coal-fired power stations dominate electricity generation in China, this has led to a very low per capita final electricity consumption, which amounted to only 407 kWh in 1990 compared with 9710 kWh in the United States and 5165 kWh in Japan in 1987 (State Statistical Bureau, 1992b; Zhang, 1991).

There are some problems facing China's power industry, some of which are discussed below:

Low unit capacity

Of thermal power capacity installed by the end of 1990, 38.8% comprised units with a capacity of 200 MW and above, 41.5% units with less than 100 MW of capacity, and 19.7% units in between 100 and 200 MW (Ministry of Energy, 1991b). Compared with a figure of over 60% in the industrialized countries, the share of large units (200 MW and above) in China is quite small (Hu *et al.*, 1990). This has led to a high net coal consumption rate of thermal power plants, which averaged 427 grams of coal equivalent (gce)/kWh in 1990, because the small-size plants are inferior to large-size ones in terms of thermal efficiency, capital costs and environmental impacts (see Table 3.6). Thus, major potential gains can be realized by installing large units in terms of economic and thermal efficiency as well as environment benefit. According to the Ministry of Energy, in the future priority will be given to constructing large-size, high-temperature and high-pressure efficient units generating 300 MW to 600 MW[3] so that by the year 2000 the average national coal consumption rate of thermal power plants will be brought down to 355 gce/kWh, and that for newly commissioned large units should not be higher than 330 gce/kWh (Ministry of Energy, 1992b).

Table 3.6   *Coal consumption rates of domestically produced thermal power plants*

Unit capacity (MW)	6	12 to 25	50 to 100	125	200	300
Net coal consumption rate (gce/kWh)	600 to 800	500 to 510	391 to 429	382 to 386	376 to 388	376 to 382

Source: Hu *et al.* (1990).

Underdevelopment of hydropower

As discussed in Section 3.2, China's hydropower potential is estimated to be the largest in the world, and its economically exploitable capacity totals 378 GW, corresponding to 1920 TWh of annual electricity production (Zhang, 1991). By the end of 1994, however, the total capacity installed of hydropower plants was only 13% of the exploitable potential, considerably less than that of the indus-

---

[3] In moving up to large units, however, the possible technological problems should be appropriately addressed. Take the UK experience in adopting 500 MW generating units as an example. Despite considerable experience in operating plants of up to 120 MW, with plant availabilities of over 80% being consistently achieved in the late 1950s and early 1960s, the rapid adoption of successively larger units led to a serious decline in plant availability which, even by 1980, had recovered only to 71% (Monopolies and Mergers Commission, 1981).

trialized countries and also below that of developing countries such as Brazil and India (see Table 3.7). Given China's abundant hydropower resources, their underdevelopment and their importance as an alternative to coal use for electricity generation, this current situation means that considerable efforts need to be devoted to speeding up hydropower exploitation in some river sections with favourable exploitation conditions. According to the Ministry of Energy, it is planned that by 2000 China's total hydropower capacity will go up to 80 GW, representing 21.1% of the economically exploitable capacity.

Table 3.7   *Hydropower exploitation in selected countries in 1990*

Country	Capacity			Electricity generation		
	Installed (GW)	Potential (GW)	I/P ratio[a] (%)	Current (TWh)	Potential (TWh)	C/P ratio[b] (%)
USA	90.10	147.25	61.2	279.8	457.1	61.2
Former USSR	64.98	315.73	20.6	223.3	1420.0	15.7
Canada	59.19	163.23	36.3	293.1	593.0	49.4
Brazil	48.75	213.00	22.9	213.4	1194.9	17.9
Japan	37.48			88.0	130.5	67.4
China	36.04	378.53	9.5	126.3	1923.3	6.6
Norway	26.60			121.6	172.0	70.7
France	24.70			69.6	72.0	96.7
Italy	19.00			31.1	65.0	47.8
India	18.34	84.00	21.8	57.8	450.0	12.8
Spain	16.70			25.7	65.6	39.2
Sweden	16.40			71.5	99.0	72.2
Switzerland	11.60			30.7	41.0	74.9
Austria	10.90			32.5	53.7	60.5

[a] I/P ratio = installed-to-potential ratio.
[b] C/P ratio = current-to-potential ratio.

Source: World Energy Herald, No. 13, July 1992 (adjusted).

Small share of cogeneration units

For energy conservation purposes, the Ministry of Energy requires all boilers at places with stable thermal loads and supplying more than 10 tons of steam per hour for more than 4000 hours per year to be converted into cogeneration. However, the development of cogeneration is hindered by factors such as high investment costs of the district heat networks, low thermal prices, and some problems in the management system. As a result, the share of cogeneration units in total thermal power capacity installed fell from 20% in 1965 to 10% in 1980 and to 9.8% in 1985. By the end of 1990, this share was only 10.9%, considerably smaller than 36% of the former Soviet Union in 1980 (Hu, 1991; Zhu, 1992).

Deficiencies of capital investment in transmission lines and distribution networks

In China power is mainly generated from coal-fired and hydropower plants. Coal reserves and hydropower resources, however, are unevenly distributed geographically, as discussed in Section 3.3. This uneven geographical distribution requires a major expansion of transmission lines and power networks to realize 'sending electricity from the west to the east as well as from the north to the south' (cf. Ministry of Energy, 1992b). But capital investment in transmission lines and power networks is insufficient in China, only accounting for about 20% of the total capital expenditures in the power industry (see Table 3.8), much less than the 65% in the industrialized countries in 1980 and also the 35% in the developing countries in 1980 (Hu, 1991). As a result, the capability of transmission lines and power networks is insufficient compared with the generating capacity installed. For example, the average annual growth rate of generating capacity installed during the Seventh Five-year Plan period was 9.6%, while the corresponding figure for 35 kilovolt and above level lines was only 6%, 3.6% lower than the former (Ministry of Energy, 1991b; Zhu, 1992). The consequences of such a mismatching between power networks and generating equipment capabilities have been the weak links in power networks, the high frequency of defects on transmission, substation and distribution equipments, and the lack of safety and stability of power networks.

Table 3.8 *Breakdown of capital expenditures for power*

Share in the total capital expenditures for power	Sixth Five-year Plan period	Seventh Five-year Plan period
Hydropower (%)	27.54	18.39
Thermal power (%)	43.19	58.25
Transmission and distribution (%)	20.94	19.29

Source: Hu (1991).

3.5 Energy conservation: an international perspective

Energy conservation is of vital importance to China, not only because it saves depletable energy resources and reduces pressure on transportation and environmental pollution, but mainly because severe shortages in energy supply have been inhibiting its economic development. In recent years, although estimates vary, China has admitted annual energy shortages of approximately 20-30 million tons of coal, 10 million tons of petroleum and 70 billion kWh of electricity. Energy shortages contribute China's claimed 25-30% underutilization of its manufacturing capacity which otherwise might be devoted to increasing a fraction of its

GNP (Liu, 1989; Zhu, 1992). It is estimated that China's energy demand in 2000 will be of the order of 1400 to 1700 Mtce, even if energy conservation is taken into account, whereas the domestic supply will be likely only to meet 1400 Mtce (Shen *et al.*, 1992; Yan, 1994). Thus, if China's development plan is to materialize, the gap has to be filled through increased efforts directed at energy conservation and enhanced energy efficiency.

Indeed, the Chinese government has been placing great emphasis on energy conservation in the past decade. A series of measures has been implemented concerning the administrative, legislative, economic and technological aspects of energy policies. As shown in Figure 3.3, great progress in decoupling its GNP growth from energy consumption has been made, with an annual growth of 9.9% for the former but 5.2% for the latter during the period 1980-95. This achievement corresponds to an income elasticity of energy consumption of 0.52,[4] an accumulated energy savings of 630 Mtce and to an annual saving rate of 4.3% (Zhang, 1997d). With regard to the breakdown of the contributions, 50% of energy savings during the period 1981-88 were attributed to structural adjustments, while strengthening energy management and technologies transformation accounted for 40%, with the remainder covered by imports of energy-intensive products (Ye, 1989).

After shedding light on the accomplishments to date of China's energy conservation efforts, this section attempts to compare energy use in China with other countries, in both monetary and physical terms. The purpose is to derive China's actual potential for longer-term energy conservation and to highlight some policy measures for promoting such a potential.

3.5.1 Direct cross-country comparison of energy use

While China has enjoyed such a great success in energy conservation, its energy use per unit of GNP is still among the highest in the world. According to the World Resources Institute (1990), China's energy intensity measured as energy consumption per unit of GNP is the highest among the ten greatest economies excluding the former USSR (see Figure 3.4), which is 5.25 times that of France, 4.42 times that of Japan, 3.80 times that of Brazil, and 1.65 times that of India.

---

[4] As shown in Table 3.9, the income elasticity of energy consumption in China is quite low by international standards. In addition to energy conservation, there are other two possible explanations for this. First, the growth of energy consumption is underestimated relative to the GNP growth. Second, quantitative restrictions have kept energy consumption from rising as would otherwise have occurred. Drawing on the analysis of rationing by Neary and Roberts (1980), the quantitative restrictions act like an implicit energy tax levied at rates varying with use and fuel. Generally speaking, households face a higher implicit tax than industrial users, and oil and natural gas are taxed at a higher rate than coal. See Hussain (1995) for a further discussion.

The energy intensity of the Chinese economy is higher than that of the industrialized countries and also that of developing countries such as Brazil and India.

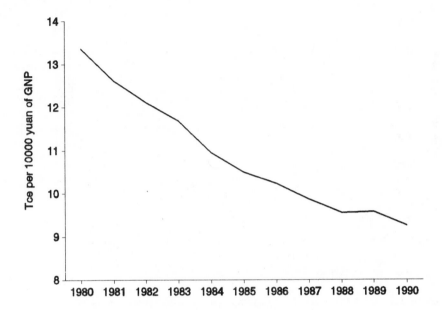

Source: Drawn based on data from the State Statistical Bureau (1992b).

Figure 3.3   *Energy intensity of the Chinese economy, 1980-90*

Table 3.9   *Growth rates of GDP and energy consumption and the income elasticity of energy consumption among different economies, 1980-90*

	Annual growth of GDP (%)	Annual growth of energy consumption (%)	Income elasticity of energy consumption
Low-income economies	6.1	5.5	0.9
Lower-middle-income economies	2.6	3.6	1.4
Upper-middle-income economies	2.4	3.6	1.5
High-income economies	3.1	1.4	0.5

Source: World Bank (1992a).

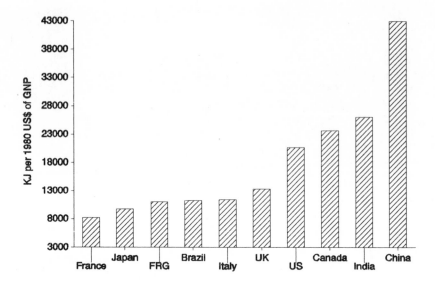

Source: Based on data from the World Resources Institute (1990).

Figure 3.4   *Energy intensity of the ten greatest economies in 1987*

This high energy intensity in China needs to be interpreted with caution however, because the differences in energy intensities of GNP between countries are not in themselves evidence of economic efficiency or inefficiency in energy use. The high energy intensity in China is partly a reflection of an unusually large share of energy-intensive industrial production in the Chinese economy on the one hand and an unusually small share of the labour-intensive service sector on the other, in comparison with other countries at its income level. For example, 40.6% of China's GDP in 1987 originated from the industry sector and 31.0% from the service sector, while the corresponding figures for India were 28.4% and 40.2% respectively (World Bank, 1992b). Moreover, the differing composition of industry affects the levels of energy intensity. Compared with 63.2% for India, China has a larger share of energy-intensive manufacturing in industry that amounted to 80.0% in 1987, thus using more energy than India per unit of industrial output, although the unit energy consumption for major industrial products in China is lower than in India (cf. TERI and INET, 1990; World Bank, 1992b). Also, energy intensity is likely to differ among countries due to differences in resource endowments and in relative resource prices, which suggests that the most economically efficient technology will differ among countries. Given that China is one of the few countries in the world that relies on coal as its major source of energy, its high energy intensity occurs partly as a result of its high proportion of coal consumption, because coal technologies are less efficient than oil/gas technologies. In addition, undervaluation of China's

GDP could also be part of the explanation. Some studies (cf. Shah and Larsen, 1992) show that China's GDP in 1987 calculated on the basis of purchasing power parities (also called real GDP) may reach 7.8 times as much as those calculated directly on the basis of exchange rate. As a result, China's energy consumption per unit of real GDP is lower than that of all above-mentioned industrialized countries, because there are not many changes in their GDP calculated on either basis. But it should be pointed out that calculating real GDP is not an easy task, particularly for the centrally planned economies. Thus, the magnitude of China's real GDP mentioned above should be considered only an illustrative figure. Certainly, it is still open to question. Nevertheless, the revised GDP does suggest that if purchasing power parities were used to measure the Chinese economy, China's energy consumption level would not appear that high.

From the preceding analysis, it thus follows that direct cross-country comparison of energy use per unit of output value should by no means be interpreted as representing the actual potential of energy conservation, given the great differences in the industrial structure and products mix among the countries selected and problems arising from using the official exchange rate. It can provide only a rough picture of relative energy intensities in selected countries.

3.5.2 Comparison of energy use in physical terms

To derive the actual potential for energy conservation in China, there is a need for a sector-by-sector comparison of energy intensity rather than for the economy as a whole. Moreover, such a comparison should be made in physical terms as much as possible.

According to China's energy balance table, industry is the dominant energy-consuming sector, accounting for 67.1% of its total final energy consumption in 1990. Within this sector, the chief energy consumers are the iron and steel industry, chemical industry, and building materials industry. The residential sector is at present the second largest user, consuming 16.8% of the total final energy use in 1990. The transportation sector in 1990 only consumed 4.7% of the total final energy use, but as much as 29.0% of the total gasoline and diesel use. Considering coal consumption, the power sector is the major consumer, with coal inputs to electricity generation in 1990 accounting for 25.8% of the total national consumption. As far as energy end-use devices are concerned, industrial boilers are the major energy users, consuming about one-third of the indigenous coal production. In what follows, the comparison will be targeted mainly at these major energy-consuming sectors and devices.

3.5.2.1 Iron and steel industry
As shown in Table 3.10, the comparable energy consumption (1.03 tce) per ton of steel produced in 1990 in China was 42% higher than the level in Italy in 1980 (Li, 1992; World Bank, 1985a).

Table 3.10   *Comparison of unit energy consumption in some energy-intensive industries*

	1980 China	1994 China	Advanced level abroad
Net coal consumption of coal-fired plants (gce/kWh)	448	413	327 (ex-USSR)
Comparable energy consumption per ton of steel (tce/t)	1.30	1.03*	0.6 (Italy)
Energy consumption per ton of synthetic ammonia (tce/t)			1.2
Large plants	1.45	1.34*	
Small plants	2.90	2.09	
Energy consumption per ton of cement clinker (kgce/t)	206.5	175.3	108.4 (Japan)

* The statistics marked with * were in 1990.

Sources: Li (1992); State Economic and Trade Commission (1996); State Statistical Bureau (1987, 1992b); World Bank (1985a).

3.5.2.2 Chemical industry
The unit energy consumption in the synthetic ammonia production differs among plants. Large-scale plants based on imported technology and using natural gas as feedstocks consumed 1.34 tce per ton of synthetic ammonia in 1990, a level similar to modern plants abroad, while small-scale ones consumed 2.09 tce per ton in 1994, about 1.75 times that of the advanced plants abroad (State Statistical Bureau, 1992b; State Economic and Trade Commission, 1996).

3.5.2.3 Building materials industry
About two-thirds of total cement production is currently in small-scale kilns (Zhou *et al.*, 1989). About 60% of these kilns are traditional vertical kilns that consume 20% more energy to produce a ton of cement than mechanized kilns (Zhai, 1993; Zhou *et al.*, 1989). The wet process, consuming twice as much energy per ton as the modern dry process, is still widely used in China's medium- and large-scale kilns, in which more than half of the cement production is produced (Zhai, 1993; Zhou *et al.*, 1989). As a result, unit energy consumption in China is one-third higher than the advanced level abroad (Zhou *et al.*, 1989; Li, 1992).

3.5.2.4 Power industry
Power generated from thermal plants accounted for 81.9% of the total electricity production of 928.1 TWh in 1994 (State Economic and Trade Commission, 1996). The average net coal consumption of coal-fired power plants amounted to 413 gce/kWh in 1994, 86 gce/kWh higher than the lowest level abroad (for example, 327 gce/kWh for the former Soviet Union in 1984) (State Statistical Bureau, 1987; State Economic and Trade Commission, 1996). This high coal consumption rate in China is to a large extent caused by low unit capacity, which has been discussed in Section 3.4.

3.5.2.5 Industrial boilers

At present, there are some 400,000 small industrial boilers across the country, about half of which have a capacity of less than 2 tons of steam per hour (t/h) (Zhai, 1993). These boilers consume 300 million tons of coal, about one-third of the national coal production (Zhai, 1993). The average thermal efficiency of these boilers is 60-70% at best, where the corresponding figure in OECD countries is 80% (Qu, 1992a; Bates and Moore, 1992).

3.5.2.6 Residential

Coal is the main fuel used for cooking in China's cities. According to surveys, in 1990, coal stoves with thermal efficiencies of only 15-30% were used by more than 60% of the total urban population, while only about 38.6% had gas stoves with double the efficiency of coal stoves (Li, 1992; Zhang, 1991). Also energy use for space heating in the cities is quite inefficient. District heating, a particularly efficient method for heat supply, was used for only 5.7% of the residential floor area in the northern cities by 1985, while about half of the floor area was heated by small coal stoves (Zhang, 1991). As a result, the average heat efficiency did not exceed 50%. In rural areas, end-use efficiencies tended to be even lower than in the cities. The average efficiency of fuel use was 14% in 1985 (Zhang, 1991).

3.5.2.7 Transportation sector

Coal-fired steam locomotives, which have almost disappeared in industrialized countries, still accounted for 29% of the total gross converted ton-kilometres in 1990, while only 18% of the total was transported by electric traction with an efficiency of two times higher than that of steam traction (Yang, 1991). As for road freight transport, mainly medium-sized, gasoline-fuelled trucks are used in China. Generally speaking, the oil consumption per vehicle-km is 20 to 30% higher than that of similar foreign trucks (Zhang, 1991). In shipping, the specific oil consumptions of diesel engines used with low- and medium-speed are 160 to 168 grams per horse-power and hour (hp.h) and 165 to 175 grams/hp.h respectively. Both figures are 25 grams/hp.h higher than that of the industrialized countries (Zhang, 1991). In addition, a number of ocean-going freight vessels are still in use after 20 years of service. China's civil aviation fleet comprises a limited number of modern aircraft with low specific fuel consumption. The petroleum pipelines are of old design with open delivery and side discharging pot. The pumps and heating systems used for transporting oil are 25-30% and 10% less efficient than those of the industrialized countries (Zhang, 1991).

The foregoing comparison of energy consumption per physical unit of output and end-use efficiencies in China with other countries clearly indicates that the efficiency in China is at the low end. Altogether, it has been estimated that energy consumption per physical unit of output in China is on average 40% higher than that of industrialized countries and that the efficiency of energy utilization in China is about 30% (Zhu, 1992; Shen *et al.*, 1992).

3.5.3 Policy discussion on promoting energy conservation

In 1980, the Chinese government initiated a policy of 'laying equal emphasis on both energy exploitation and conservation, and priority given to the latter in the near future' (Li, 1992). In practice, however, this policy has become exploitation-dominated. The following statistics support this view.

As shown in Table 3.11, total fixed asset investment in the energy sector by state-owned enterprises during the period 1981-90 reached 435.5 billion yuan. Meanwhile, technical updating investment in energy conservation, combined with capital construction investment in energy conservation, amounted to 27.2 billion yuan, accounting for only 6.2% of the total fixed asset investment in the energy sector. This means that 93.8% of energy investment funds were allocated for expanding energy supply.[5] Given such imbalance between investment in expanding energy supply and investment in energy conservation, how can it be explained that equal emphasis is placed on both energy exploitation and conservation?

Table 3.11 *Energy investment by state-owned enterprises (100 million yuan)*

	1981-85	1986-90	1981-90
Total fixed asset investment in the energy sector	1171.08	3184.37	4355.45
Total energy conservation investment	112.27	159.61	271.88
Capital construction investment in energy conservation	59.26	113.09	172.35
Technical updating investment in energy conservation	53.01	46.52	99.53

Sources: State Statistical Bureau (1992b); *Handbook of Rational Utilization of Resources*, China Science and Technology Publishing Press, Beijing, 1991.

The lack of investment in energy conservation is explained briefly below. Economic reforms over the past 15 years in China have devolved control over resources and decision making to local governments as well as enterprises. While the central government still undertakes about 70% of budgetary financing primarily for capital construction investment, most technical updating investment is now under the control of local governments and enterprises, which rely on bank loans to finance their projects (cf. Singh, 1992). But given the current subsidies for energy consumption that vary with fuel (see Table 3.12 for an

---

[5] Some of the funds could have been used for employing more advanced technologies. This could help to minimize the energy use in the course of energy production itself but certainly cannot avoid or diminish inefficient energy use on the demand side. Moreover, it is generally thought that the supply-oriented approach to energy development is more expensive than the demand side management aimed at energy conservation. Thus, heavy reliance on the continued expansion of energy production may represent a serious misallocation of resources in any *bona fide* attempt to promote energy conservation.

economic measure of the subsidies),[6] local governments and enterprises have little incentive to invest in energy conservation.

Table 3.12 *Subsidies for fossil fuel consumption in selected countries (measured by the ratio of domestic prices to world prices)*

Country	Coal	Oil	Natural Gas
China	0.84 (1989)	0.48 (1985)	0.40 (1986)
Former USSR	0.10 (1992)	0.05 (1992)	0.07 (1992)
Poland	0.30 (1990)	0.68 (1990)	0.50 (1990)
India	0.86 (1991)	0.47 (1990)	-
South Africa	0.50 (1991)	-	-
Czechoslovakia	0.30 (1990)	0.22 (1990)	-
Mexico	-	0.54 (1990)	-

Source: Larsen and Shah (1992).

I shall return to the consequences of insufficient energy conservation investment later, but I now shall discuss briefly China's pricing system, because it is of great importance to encouraging energy conservation investment.

Before the post-1978 economic reforms, China's economic management structure was modelled principally on that of the former Soviet Union, an essential feature of which was the adoption of a unified state pricing system. Under the pricing system, the state-set prices of goods, including those of energy, reflected neither the production costs nor the influence of market forces. The structure of state-set prices was also irrational: the same types of goods were set at the same prices regardless of their qualities, thus resulting in the underpricing and undersupply of goods of high quality. Over a too long period, the pricing system remained unchanged so that its inflexible and restrictive nature became increasingly apparent. Thus, the outdated pricing system had to be reformed.

In 1984, the government allowed state-owned enterprises to sell their above-quota or surplus at prices within a 20 percent range above the state-set prices (Wu and Zhao, 1987). In February 1985, the 20% limit was removed and prices for surplus could be negotiated freely between buyers and sellers (Wu and Zhao, 1987). At that point, the dual pricing system was formally instituted. Such a pricing system is generally considered a positive, cautious step towards a full market price. Moreover, it is widely thought that introducing the dual pricing system can, among other purposes, encourage material and energy conservation

---

[6] The subsidies for energy consumption also vary with use. Albouy (1991), for instance, shows that industrial users enjoy much higher subsidies for coal use than households. The latter are also charged the highest electricity tariffs.

and improved management, thus introducing economic efficiency in the use of resources.[7]

Table 3.13 presents some data on plan and market prices as well as data on plan allocations from a survey of 17 provincial markets.

Table 3.13   *Ratio of market price to plan price, and percentage of plan allocation of selected goods by volume and value, March 1989*

Material	Ratio of market price to plan price	Percentage of plan allocation by volume	Percentage of plan allocation by value
Crude oil	3.13	80	56
Heavy oil	2.60	41	13
Copper	2.50	17	7
Coal	2.49	46	21
Gasoline	2.25	64	44
Aluminium	2.24	28	15
Fertilizer	2.23	39	26
Timber	2.12	22	12
Diesel fuel	2.05	55	36
Steel products	2.05	30	19
Electric power	1.89	75	60
Nitric acid	1.82	40	20
Soda ash	1.81	40	28
Plate glass	1.63	41	29
Aluminium products	1.63	6	4
Caustic soda	1.60	47	24
Kerosene	1.60	73	67
Copper products	1.49	8	5
Cement	1.36	16	11
Iron ore	1.33	78	74
Pesticide	1.33	62	54
Sulphuric acid	1.30	40	32
Crude salt	1.23	86	83
Pig iron	1.10	47	42

Source: China Price, September 1990 (adjusted).

---

[7] See Wu and Zhao (1987) and Singh (1992) for general discussion on advantages and disadvantages of the dual pricing system and Albouy (1991) for the investigation of its impact on coal.

Of particular interest is the continued importance of the plan in the allocation of energy goods, particularly crude oil and electricity. This means that state-owned enterprises still receive allocation for part of their energy inputs at the state plan prices. As shown in Table 3.13, however, the state-set plan prices of energy goods are kept much lower than their market prices. As a result, these enterprises have weak incentives for investment in energy conservation.

Let us now turn to the consequence of insufficient energy conservation investment. Such insufficiency has led to a reduction in energy savings attributed to technical updating. As shown in Table 3.14, the energy saving costs defined to be the investment costs per ton of coal equivalent saved were reported to rise to 523 yuan during the Seventh Five-year Plan period (1986-90), 63.4% higher than the average 320 yuan during the Sixth Five-year Plan period (1981-85). Total energy conservation investment should keep the same pace of growth as the energy saving cost, if the energy savings during the period 1986-90 were to match those of the 1981-85 span. During the period 1986-90, however, total energy conservation investment had increased by 4.7 billion yuan, only 42.2% more than the investment during the period 1981-85 (see Table 3.11). As a result, energy savings attributed to technical updating during the period 1986-90 amounted to only 30.5 Mtce, 4.6 Mtce less than that during the period 1981-85.

Table 3.14   *Energy savings attributed to technical updating and their unit costs*

	Energy savings (Mtce)		Unit costs (yuan/tce saved)	
	81-85	86-90	81-85	86-90
Capital construction investment in energy conservation	13.90	20.95	426.3	539.8
Technical updating investment in energy conservation	21.20	9.59	250.0	485.3
Total energy conservation investment	35.10	30.54	319.8	522.8

Source: Handbook of Rational Utilization of Resources, China Science and Technology Publishing Press, Beijing, 1991.

It is therefore clear that, in order to encourage future energy conservation investment and hence efficiency gains, current subsidies for energy consumption in China should be eliminated. Such a reform of energy pricing will give economic incentives for energy conservation. Also, consideration may be given to appropriate control over the growth of China's energy supply in order to put pressure of energy conservation on the demand side.

3.6   Analysis of historical CO$_2$ emissions

In this section, I first examine the historical evolution of CO$_2$ emissions over the past decade. Then I analyse the historical contributions of inter-fuel switching, energy conservation, economic growth and population expansion to CO_2 emissions.

3.6.1 Historical evolution of CO_2 emissions

On the basis of fossil fuel consumption, the corresponding CO_2 emissions have been calculated by using the CO_2 emission coefficients given in Table 3.15. These coefficients are measured in tons of carbon per ton of coal equivalent (tC/tce) and generally considered suitable for China.

Table 3.15   *CO_2 emission coefficients for China*

Fuels	tC/tce
Coal	0.651
Oil	0.543
Natural gas	0.404
Hydropower, Nuclear power and Renewables	0

Source: Energy Research Institute (1991).

Table 3.16 shows total CO_2 emissions from fossil fuels in China over the period 1980-90. It is clear that total CO_2 emissions in China rose from 358.60 MtC (Million tons of Carbon) in 1980 to 586.87 MtC in 1990, with an average annual growth rate of 5%. This means that China ranks third in global CO_2 emissions, behind the US and the former Soviet Union, and that if the Soviet emissions are distributed over the new independent republics, China is second. But on a per capita basis, China's CO_2 emissions of 0.5 tC in 1990 were very low, only about half the world average.

The breakdown of CO_2 emissions by fuel is shown in Figure 3.5. Because of the coal-dominant structure of energy consumption, it is not surprising that coal predominates, accounting for 83.4% of total emissions in 1990.

3.6.2 The contributions to CO_2 emissions growth

CO_2 emissions can be subdivided as follows (see, for example, Ogawa (1991), and Dean and Hoeller (1992) for different subdivisions):[8]

---

[8] This is a concrete form of the so-called Ehrlich equation $I = PAT$, where I represents the adverse environmental impact, P is the population, A is the consumption per capita, and T is the amount of resources required by environmentally damaging technology for producing one unit of consumption (cf. Ehrlich and Ehrlich, 1990). It is used as a proxy for a determinant of environmental impact.

$$C = \left(\frac{C}{FEC}\right) \cdot \left(\frac{FEC}{TEC}\right) \cdot \left(\frac{TEC}{GDP}\right) \cdot \left(\frac{GDP}{POP}\right) \cdot POP$$

where C is the amount of CO$_2$ emissions, FEC is the total carbon-based fossil fuel consumption, TEC is the total commercial energy consumption, GDP is the Gross Domestic Products, and POP is the population.

Table 3.16 *CO$_2$ emissions and their determining factors[a]*

Year	C (MtC)	C/FEC (tC/tce)	FEC/TEC	TEC/GDP [b]	GDP/POP (yuan [c])	POP (million)
1980	358.60	0.62	0.96	1.03	594.97	987.05
1981	352.63	0.62	0.95	0.97	613.31	1000.72
1982	367.82	0.62	0.95	0.93	656.19	1016.54
1983	390.39	0.62	0.95	0.90	713.06	1030.08
1984	421.41	0.63	0.95	0.84	805.47	1043.57
1985	456.58	0.63	0.95	0.81	891.44	1058.51
1986	482.01	0.63	0.95	0.79	950.40	1075.07
1987	517.32	0.63	0.95	0.76	1036.63	1093.00
1988	554.98	0.63	0.95	0.75	1119.71	1110.26
1989	577.37	0.63	0.95	0.75	1140.73	1127.04
1990	586.87	0.63	0.95	0.73	1174.75	1143.33

[a] Symbols are explained in Section 3.6.2.
[b] Measured in metric tons of coal equivalent per thousand Chinese yuan at 1987 prices.
[c] At 1987 prices.

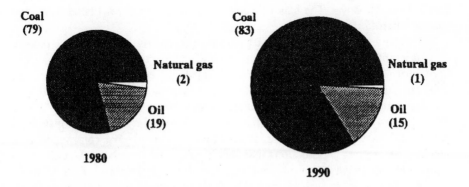

Figure 3.5 *Sources of CO$_2$ emissions in 1980 and 1990*

Taking logs and differences over time yields:

$$\Delta \log C = \Delta \log(C/FEC) + \Delta \log(FEC/TEC) + \Delta \log(TEC/GDP)$$
$$+ \Delta \log(GDP/POP) + \Delta \log(POP)$$

The first term on the right-hand side of the identity shows the effect of changes in the composition of carbon-based fossil fuels on emissions, and the second term indicates the contribution of the penetration of carbon-free fuels (1-FEC/TEC) to a reduction in emissions (if the share of carbon-free fuels (1-FEC/TEC) is increased, the CO_2 emissions can be effectively reduced). These two terms therefore capture the contribution of inter-fuel substitution to the changes in emissions, as explained below: fuels vary considerably in their relative CO_2 emissions. Specific CO_2 emission from coal burning is 1.6 times that from natural gas and 1.2 times that from oil. Hydropower, nuclear energy and renewables do not produce CO_2 emissions. In this regard, increased use of carbon-free energy sources, along with substitution of natural gas for the more pollution-producing coal and oil, would clearly reduce CO_2 emissions.

The third term shows the effect of changes in the aggregate energy intensity on emissions, and the last two terms show the effect on emissions due to growth in income per capita and population respectively. Needless to say, this identity is in a form suitable for analysing the historical contributions of inter-fuel switching, energy conservation, economic growth and population expansion to CO_2 emissions by examining the relevant time-series data.

Table 3.17 shows the results of this analysis for the period 1980-90, based on data given in Table 3.16. It quantifies the historical contribution to CO_2 emissions each factor has made. Population data and commercial energy consumption of various types have been taken from the State Statistical Bureau (1992a; 1992b), and GDP values have been derived from the World Bank (1992b). The corresponding CO_2 emissions associated with the fossil fuel consumption have been calculated above. The data of Table 3.16 are presented in Figure 3.6, after normalization to the year 1980.

The results in Table 3.17 and Figure 3.6 clearly indicate the relative importance of each factor in terms of its contribution to CO_2 emissions growth. Given that China had been the most rapidly expanding economy over that decade, it is not surprising that economic growth measured in per capita GDP was overwhelming. This factor alone resulted in an increase of 314.67 MtC. During the corresponding period, China experienced a lower rate of population growth through the strict family planning programmes, which in turn contributed to a smaller increase in CO_2 emissions than would otherwise have been the case. As a result, the population expansion was responsible for an increase of 68.27 MtC. The contribution to an increase in emissions is considered to be modest given its population size. Also the change in fossil fuel mix contributed to an increase in emissions (5.33 MtC), but its role was very limited because the share of coal use in total commercial energy consumption was slightly increased during the period.

Table 3.17 *Breakdown of the contributions to CO$_2$ emissions growth during the period 1980-90 (MtC)*

Due to change in fossil fuel carbon intensity	Due to penetration of carbon free fuel	Due to change in energy intensity	Due to economic growth	Due to population expansion	Total change in CO$_2$ emissions
+5.33	-5.33	-154.67	+314.67	+68.27	+228.27

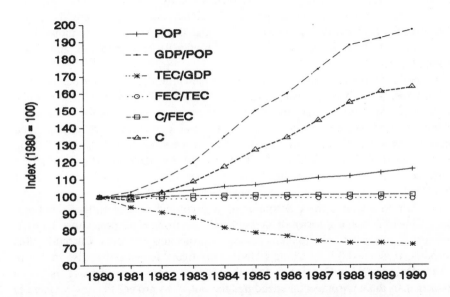

Figure 3.6 *Contributions to CO$_2$ emissions in China, 1980-90*

By contrast, reduction in energy intensity tended to push CO$_2$ emissions down. During the period 1980-90, great progress in decoupling China's GNP growth from energy consumption had been made, with an annual growth of 8.97% for the former but 5.06% for the latter. This achievement corresponds to an income elasticity of energy consumption of 0.56, an accumulated energy savings of 280 Mtce and to an annual saving rate of 3.6%, higher than the planned target of 3% (Shen *et al.*, 1992). As a result, a reduction of 154.67 MtC was achieved. Also the penetration of carbon-free fuels contributed to a small reduction in CO$_2$ emissions (-5.33 MtC). This is mainly due to the underdevelopment of hydropower, and partly because no nuclear power stations were commissioned during that period.

3.7   Environmental challenges for the Chinese energy system

Environmental problems related to energy use in China reflect the structure of energy supply and consumption. As discussed in Section 3.3, because heavy reliance is placed on coal and in particular the extensive direct use of coal, the environmental effects of energy use lie mostly with coal burning. In 1993, the total emissions of SO_2 and smoke reached 17.95 million tons and 14.16 million tons respectively, about 90% and 70% of which respectively came from coal burning. Consequently, the average daily concentration of total suspended particulates (TSP) in the air in northern cities was as high as 407 $\mu g/m^3$ and that of SO_2 was 100 $\mu g/m^3$. SO_2 concentrations in northern cities during winter quite frequently exceed the government standards. The same applies to TSP concentrations. Furthermore, the situation is gradually deteriorating in some cities. The south of China, particularly Sichuan and Guizhou, has been seriously affected by acid rain, to which oxides of sulphur and nitrogen are the major contributors. Nationwide, the area of farmland that had been contaminated by acid rain in 1993 amounted to 5.3 million hectares (State Economic and Trade Commission, 1995).

Air pollution poses an immediate threat to human health and a serious ecological damage. It has been estimated that annual losses from air pollution during the Sixth Five-year period were about 12.5 billion yuan. When losses from water pollution, land occupation by solid wastes, and pesticides were added, total annual losses during the corresponding period were estimated at 38.2 billion yuan. This constituted 6.75% of China's GNP for that period, whereas the corresponding figure for industrialized countries was estimated to be 1-5% of their GNP (Guo *et al.*, 1990; Pearce and Warford, 1993).

In order to alleviate the environmental impacts and control further deterioration, China has taken a series of measures for environmental protection. Increasing funds are spent on the implementation of protection measures. Currently, this spending is about 0.8% of China's GNP. This figure is planned to rise to 1% in 2000 (Qu, 1992b).[9] Given the fact that the environmental damage is much greater than the environmental spending, the policy to protect the environment is more than justified.

As energy consumption in general and coal use in particular is the major source of air pollution from the perspective of environmental protection, it is thus not surprising that efforts to combat air pollution are targeted at the energy sector, particularly at coal use. Some of the energy-related measures that are being, and will continue to be, implemented include the following.

---

[9] In an interview with China Daily (16 December 1995), Chairman of the Environmental and Resource Protection Committee of China National People's Congress said that China is planning to raise the share of its environmental investment to 1.5% of China's GDP by 2000.

Increasing proportion of raw coal washed The ash content of raw coal from state-owned mines is about 30% by weight (Wu, 1995). With coal washing, ash content drops by 39% to 18.2%. Among steam coal operations, the ash content of raw coal averages 27.4%, while the ash content of washed coal drops to 21.5%, a 21.5% decrease (Ward *et al.*, 1994). By reducing the ash content of coal, coal washing will reduce the amount of coal to be transported over a long distance to the load centres, improve the efficiency of coal use, and reduce particulate emissions. Energy efficiency studies show that power plant efficiency is decreased by 0.2% for each percentage increase in ash content. Given average ash reductions of 5-7% attributed to washing for steam coal and even higher reductions for coking coal, the overall efficiency savings could be substantial. A study based on coal washing plants in Datong (Shanxi province) and Xuzhou (Jiangsu province) has yielded the internal rates of return of above 18.8%, although the returns are sensitive to the prices of both raw and washed coal and actual production levels (cf. Ward *et al.*, 1994). Because coal washing is economic and is also justified on grounds of environmental protection, the Chinese government is determined to continue the development of coal washing and to increase its capacities. In the future, mines and washing plants are required to be planned and completed simultaneously, and mines without washing plants will not be approved. As such, the proportion of raw coal washed is planned to be brought up to 40% in 2000 from 20% in 1993 (State Economic and Trade Commission, 1995; Wu, 1995).

Retrofitting and replacement of small inefficient industrial boilers As discussed in Section 3.5, currently, some 400,000 small inefficient industrial boilers across the country consume about one-third of the national coal production. Thus, there is much room for coal savings and reductions in emissions of pollutants. For this purpose, the Ministry of Energy requires conversion to cogeneration of all boilers at places with stable thermal loads and supplying more than 10 tons of steam per hour for more than 4000 hours per year.

Substitution of direct burning of coal by electricity through development of large-size, high-temperature and high-pressure efficient coal-fired power plants In 1990, only 25.8% of national coal output was used for electricity generation, and direct use of coal accounted for as much as 68% of the total (State Statistical Bureau, 1992b). Thus, in order to alleviate the environmental impact of coal use, more coal should be transformed to such high-quality energy as electricity. It is expected that by 2000 the proportion of coal used for electricity generation will rise to about 35%. Moreover, in the construction of coal-fired power plants, in the future priority should be given to installation of large-size, high-temperature and high-pressure efficient units generating 300 MW to 600 MW - see Section 3.4 for a further discussion.

Speeding up hydropower exploitation This has been given extensive discussion in Section 3.4.

Popularizing domestic use of coal briquette Coal stoves using coal briquette can reduce coal consumption by 20-30%, CO emissions by 70-80%, and SO$_2$ emissions by 40-50% if sulphur-fixing additives are added to the briquette (Zhao and Wu, 1992). In 1990, the production of coal briquette for domestic use reached 33 million tons, accounting for 20% of the total domestic coal use. Given that coal consumption still accounts for about 80% of total domestic energy use, popularizing domestic use of coal briquette can certainly realize major potential gains in terms of thermal efficiency and environment benefit.

Increased penetration of town gas into urban households Town gas is one of the only long-term options for displacing direct coal burning in China's residential sector. In 1980, an urban population of 11 million had access to town gas for cooking. After the great efforts over the following 14 years, the corresponding figure for 1994 rose to 104.22 million, 9.5 times that of 1980. Consequently, the urban gasification rate was 61.7% (State Economic and Trade Commission, 1996). Clearly, the increased penetration of town gas can reduce urban pollution. In general, however, a widespread substitution of town gas for coal is limited by the lack of supplies.

Expanding district heating systems Over the past 14 years, investment has been made in the construction of heat supply networks in many cities. By 1994, such heating networks had connected a floor area of 505.97 million m^2, 297 times that of 1980 (Qu, 1992a; State Economic and Trade Commission, 1996). Because district heating is a particularly efficient method for space heat supply, further expansion is expected to take place.

In addition, efforts are being made for research and development of environmentally sound coal technologies, including circulating fluidized bed combustion boilers, coal-water slurry, and coal gasification combined cycle.

It should be pointed out that success in the implementation of these 'no-regrets' measures will largely depend on the extent to which a reform of energy pricing will be carried out. Given the current subsidies for energy consumption and without change, it is unlikely that the required outcome will be achieved.

Although all these measures can alleviate the environmental impact of coal use, curbing global CO_2 emissions calls for limitations to the overall consumption of coal, because specific CO_2 emission from coal burning is nearly twice that from natural gas and one and a half times that from oil. This will present an additional challenge to China's energy development strategies.

At present, the reduction of CO_2 emissions is not at the top of Chinese government agenda, although the bulk of the above-mentioned measures taken to combat air pollution also reduce CO_2 emissions. The Chinese authorities know that their own CO_2 emissions, though high in relation to population size and energy use, so far have still been well below the world average level on a per capita basis, because of the low level of development of the Chinese economy. They are also aware that China is bound to rely mainly on coal as fuel in the foreseeable future. Given this prospect, it is generally thought that options for China to limit CO_2 emissions are high-cost supply substitutes and price-induced energy conservation in line with rapid economic growth, and thus that China would be the region hardest hit by carbon constraints (Manne and Richels, 1991a). Under these circumstances, the Chinese authorities have claimed that ignoring the past build-up by the industrialized countries and simply asking for equal percentage of CO$_2$ reductions on current emissions level would seriously harm China's economic development and improvement of living standards and thus cannot be accepted as fair (Bai, 1991). This is generally in line with the current attitudes of the developing countries towards greenhouse gas controls.[10]

Of course, this is not to justify the inaction by China; the Chinese government has indicated its willingness to take all possible measures to limit its own CO$_2$ emissions. Until now, the Chinese government has ratified the United Nations Framework Convention on Climate Change and China's Agenda 21, the latter of which serves as a white paper on China's population, environment and development in the 21st century. Although no concrete commitment has been made yet to limit CO$_2$ emissions on the Agenda, it is conceivable that China's foreseeable positive and cooperative actions will be reflected in a broad range of measures to slow down, to a large extent, the growth of per capita CO$_2$ emissions. With such efforts, China may keep its per capita CO$_2$ emissions well below the world average level without jeopardizing its economic development. This can be seen as a reasonable and achievable target for China.

3.8   China's potential importance as a source of CO$_2$ emissions: observations from global studies

Because of the global characteristics of climate change and China's potential importance as a source of CO$_2$ emissions, there exist, though relatively few, global models that cover various political-economic regions and that treat China as a separate region. Global models in this tradition include the well-known GLOBAL 2100 (Manne and Richels, 1991a, 1991b, 1992; Manne, 1992) and GREEN (Burniaux *et al.*, 1992; Martin *et al.*, 1992; Martins *et al.*, 1992; OECD, 1993a).[11] These models have been used to predict the future levels of CO$_2$ emissions in the absence of specific policies to limit CO$_2$ emissions and to

---

[10] According to the Beijing Declaration adopted at the Developing Countries' Ministerial Conference on Environment and Development held in Beijing on 18-19 June 1991, the responsibility for greenhouse gas emissions should be viewed both in terms of historical and cumulative terms, and in terms of current emissions. Consequently, the industrialized countries must take immediate action, with binding emission targets, while the developing countries can not be expected to accept any obligations in the near future (People's Daily (Overseas Edition), 20 June 1991).

[11] There are a number of widely cited global studies that are based on the Whalley-Wigle model (cf. Whalley, 1991; Whalley and Wigle, 1991a, 1991b; Pezzey, 1992; Piggott *et al.*, 1992). Because the global model does not treat China separately, however, those studies are not touched upon in this section.

analyse the economic impacts of compliance with CO_2 emission limits on a global scale. The main findings arising from these models have already been presented in Section 2.5. Therefore, only some important observations particularly related to China are mentioned here. Prior to this, I first describe these global models briefly, because the results obtained are frequently cited in this book and serve as a basis for comparison with the case study for China in Chapter 7.

So far, the most complete global model within an optimization framework in terms of modelling the energy sector and its feedback to aggregate output has been that of Manne and Richels. The so-called GLOBAL 2100 model, with a dynamic nonlinear optimization framework of the maximization of discounted utility, is based on parallel independent computations for five major geopolitical regions: the US, other OECD nations, the former USSR and Eastern Europe, China and the rest of world. The model has a rich treatment of the energy sector but a highly aggregated description of the economy. It can be run as far into the future as the year 2100, in eleven steps of ten-year intervals. Using this model, Manne and Richels (1991a, 1992) and Manne (1992) analyse the abatement costs under alternative CO_2 emission limits, and carbon tax rates necessary for achieving large CO_2 reductions and the feedback impacts of rising energy costs. Manne and Richels (1991b, 1992) also quantify the potential for international trade in CO_2 emission permits based on an allocation rule under which the carbon permits are distributed among regions in proportion to their 1990 levels of emissions initially (the year 2000) and in proportion to the 1990 level of population at the end of the planning horizon (the year 2100).

The OECD Secretariat (Burniaux *et al.*, 1992) has developed a multi-sector, multi-region, dynamic CGE (Computable General Equilibrium) model used for evaluating the economic costs of international agreements to curb global CO_2 emissions. The model is referred to as GREEN (GeneRal Equilibrium ENvironmental model). It has full clearing markets and is made up of twelve regions. So far it has been the most complete global CGE model in terms of fuel, regional and sectoral disaggregations and the modelling of backstop technologies. Currently, GREEN is being simulated over the period 1985-2050, in five steps of five-year intervals up to 2010 and two further steps of twenty-year intervals. In each region, the model is calibrated on exogenous growth rates of GDP and population and on neutral technical progress in energy use. Given the recursive structure of the model, the evolution over time of the economy is described as a sequence of single-period static temporary equilibria. Using GREEN, Martin *et al.* (1992) and Martins *et al.* (1992) simulate the economy-wide impacts of a variety of international agreements on curbing CO_2 emissions, including: 1) a so-called Toronto-type agreement under which it is assumed that there will be a 20% cut in CO_2 emissions in the industrialized countries and a 50% rise in the developing countries relative to their 1990 levels by 2010 respectively, with emissions in each region being stabilized thereafter; 2) a Toronto-type agreement with trade in emission permits, which means that trade in carbon emission permits among regions is allowed, while imposing the same CO_2 emission limits on each region as the first; and 3) compliance with the Toronto-type agreement only in the industrialized countries. Moreover, the robustness of the results with respect to some key parameters of GREEN is assessed.

Now let us turn to the results of these studies. First, look at the baseline projections. Over the period 1990-2050, GREEN shows a rapid growth of CO$_2$ emissions for China, with an average annual growth rate of 3.7% for that period (Martin *et al.*, 1992), whereas the estimate by Manne (1992) puts the growth at 2.3%. This difference in growth rates for CO$_2$ emissions would lead to differences in the predictions for the timing of doubling China's CO$_2$ emissions and its share in the global CO$_2$ emissions. In GREEN, a doubling of China's CO$_2$ emissions would occur as early as in 2007 or so, while this would take place in 2025 in the study of Manne (1992). In terms of regional contributions in 2050, GREEN suggests that China would become the single most important CO$_2$-emitter, with its share rising from 9.5% in 1985 to 29% in 2050 (Martin *et al.*, 1992). This would leave other countries and regions far behind China. By contrast, the portion attributable to China would be estimated to be 17% in 2050 in the study of Manne (1992).[12]

When it comes to analysing the economic impacts of compliance with CO$_2$ emission limits, the study of Manne and Richels (1991a) assumes that China would be allowed to double its carbon emissions relative to the 1990 level by 2100, while CO$_2$ emissions in the industrialized countries would be restricted to 80% of their 1990 levels. Even so, in terms of foregone GDP, China would be the region hardest hit by the carbon constraint. Manne and Richels (1990) show that China would face annual GDP losses of over 10% by the latter half of the next century. Even if the carbon emissions in China were relaxed to a quadrupling, its output losses would be considerable, still amounting to 7% of GDP by 2100. As for GREEN, China would also record a welfare loss of 9% in 2050 under the Toronto-type agreement (Martin *et al.*, 1992).

Despite great variation in the magnitude of loss estimates, there are similarities among the results of these global studies. All of them clearly indicate that China would be recorded among the severest sufferers from the global CO$_2$ emission limits, if compared with an unrestricted business-as-usual scenario. The highest costs induced for compliance with the emission targets reflect that the major alternatives available for China are high-cost supply substitutes and price-induced energy conservation in line with rapid growth in GDP.

Global models are of wider regional scope and are thus able to give insight into the regional effects of different types of international agreements on international

---

[12] The baseline assumptions underlying the study of Manne (1992) are based on the guidelines laid down for the OECD model comparisons project (Dean and Hoeller, 1992). The assumptions are different from those used by Manne and Richels (1991a). This leads to differences in the baseline projections for China's share in global CO$_2$ emissions. The study of Manne (1992) shows that by 2100 the portion attributable to China is 28%, while the corresponding figure projected by Manne and Richels (1991a) is 22%.

trade and welfare consequences. In the existing global models, however, the developing countries are typically treated in a simplistic fashion or aggregated together because of lack of appropriate data and institutional information or computational limits on model size (Weyant, 1993). Given the fact that the contributions of a number of the developing countries to global CO_2 emissions, which are high already, are expected to grow significantly, there has been growing recognition that for such countries with large CO_2 emissions as China, a country-specific model should be developed. The single-country (China) model allows for a more detailed and reliable analysis than the existing global models in terms of sectoral scope and energy sources. Such an analysis on a national level is useful to broaden the picture painted by global models. Rather than competing or substituting, however, these two approaches can serve to check or even complement each other, thus illustrating how the two approaches are able to shed light on different aspects of the same problem. Indeed, given the fact that CO_2 is a uniformly mixed pollutant, that is, one ton of carbon emitted anywhere on earth has the same effect as one ton emitted somewhere else, it does not matter whether CO_2 emissions are reduced in the United States or in China. What matters is whether we are able to reduce the emissions effectively on a global scale. One unit of US dollars spent in a cost-efficient strategy of the US may provide less gain in terms of the amount of CO_2 emissions reduced than the same amount spent in a Chinese cost-efficient strategy, assuming that the total world investment in carbon programmes remains the same. In this regard, the country-oriented studies are only of limited use if they are not compared with each other within a global framework.

In the next chapter alternative economic modelling approaches to cost estimates for limiting CO_2 emissions will be discussed. Such a discussion will show the theoretical rationale for choosing a computable general equilibrium (CGE) approach to the macroeconomic analysis of CO_2 emission limits in China and for linking a CGE model of the Chinese economy with a power planning model of China's electricity sector.

4 Economic Modelling Approaches to Cost Estimates for Limiting CO_2 Emissions[1]

4.1 Introduction

Given the uncertainties surrounding the effects of climate change and the difficulties in quantifying the damages avoided, most studies are confined to emission abatement costs and abatement strategies in isolation from their expected benefits, although this may lead to a policy bias towards inaction.[2] Moreover, with carbon dioxide (CO_2) thought to be responsible for half of the present global warming and all GHG, with the exception of CFCs, associated to a greater or lesser extent with the combustion and/or production of fossil fuels, empirical studies have placed emphasis on cost estimates for reducing CO_2 emissions from fossil fuels combustion.[3]

Just as global warming estimates are based on the conventional benchmark of a doubling of atmospheric CO_2 concentration from its pre-industrial level, cost estimates for limiting CO_2 emissions also require some common but arbitrary objective in order to be comparable. Most estimates take as the target a reduction of emissions to either a specified fraction of what they would have been in the absence of control, or some fixed proportion of the emissions in, say, 1990 (cf. Schelling, 1992). The costs of CO_2 abatement are then estimated through the *ad hoc* approach, dynamic optimization approach, input-output and macroeconomic approaches, computable general equilibrium (CGE) approach, and hybrid approach respectively. Without going into too much detail, this chapter attempts to highlight the relative strengths and weaknesses of these different economic

---

[1] This chapter is a much revised and expanded version of an article in *International Journal of Environment and Pollution* (Zhang, 1995a), the Chinese version of which appeared in *Bulletin of Energy Policy Research* (Zhang, 1995c).

[2] If governments decide to pursue mitigation policies, they would probably do so on the basis of anticipated benefits rather than costs. With ample estimates on the cost side and few estimates on the benefit side of global warming mitigation, there could be a policy bias towards inaction (Cline, 1991).

[3] For a survey of empirical studies on the estimates of economic costs of controlling GHG emissions, see, for example, Boero *et al.* (1991), Dean and Hoeller (1992), Hoeller *et al.* (1991), Nordhaus (1991a), Grubb and Edmonds *et al.* (1993), and Gaskins and Weyant (1995).

approaches. Its purpose is to illustrate how these different economic approaches are able to shed light on different aspects of cost estimates for the control of CO_2 emissions and, at the same time, to show the rationale for choosing a CGE approach for the macroeconomic analysis of CO_2 emission limits for China and linking such a CGE model of the Chinese economy with a power planning model of China's electricity sector. Moreover, some conclusions with respect to the applicability of each approach are drawn.

4.2   The *ad hoc* approach

The *ad hoc* approach usually comes down to a comparison of a limited number of CO_2 abatement options. It has been exemplified by pairwise comparison of nuclear and efficiency abatement strategies (Keepin and Kats, 1988) and by a comparative analysis of seventeen different abatement options (Jackson, 1991). Investigating and comparing the cost of specific low-CO_2 technologies, for example, CO_2 scrubbing and substitution of methane for oil and gas, fall into this category. The main purpose of such a comparison is to identify cost-efficient technologies for achieving the specified goals. Moreover, it allows for ranking the options examined in terms of their cost-effectiveness and hence prioritizes investments in greenhouse abatement. For instance, Keepin and Kats (1988) show that, in the USA, improving electric efficiency is nearly seven times cheaper than switching to nuclear power in order to achieve a given amount of CO_2 reductions. Thus revitalizing nuclear power would be a relatively expensive and ineffective response, thereby representing a serious misallocation of resources in any *bona fide* attempt to mitigate global warming. This conclusion also holds for Jackson's study. In Section 8.6, this approach will be used to compare 15 types of power plants in China in terms of both the levelized cost of generation and the marginal cost of CO_2 reduction.

Given that the costs and timing of alternative options to limit CO_2 emissions differ, the *ad hoc* approach, which is based on individual evaluation of each option, is not suited for inferring the most cost-efficient mix and scale of abatement technologies. Moreover, this approach ignores the transaction costs. In the next sections, I will focus on the dynamic optimization, traditional economic, computable general equilibrium and hybrid approaches, through which a large number of abatement technologies and activities can be examined in order to identify an optimal mix of technological options and/or to analyse their economy-wide impacts.

4.3   Dynamic optimization approach

Dynamic optimization models[4] commonly used in the CO$_2$ context fall into two broad categories. The first category refers to energy-sector optimization models (cf. Beaver, 1993). Just as the name implies, energy-sector optimization models, including the widely-used MARKAL,[5] focus solely on the energy sector and often have an explicit, detailed description of an extensive array of energy demand and supply technologies and fuels. Upon the representation of these technologies and fuels in a 'shopping list' in terms of their technical, economic and environmental characteristics, optimization models simulate the competition among fuels and technologies, and choose the most cost-efficient mix of technologies and fuels to meet the various exogenously-determined energy demands and to comply with the emission limits if the minimization of discounted cost over the entire planning horizon is chosen as the objective function.

Energy-sector optimization models are often of an intertemporal structure, and thus allow for interactions between periods. This makes models of this type very useful for assessing the potential of new technologies, especially given the uncertain parameters characterizing these new technologies (cf. Bergman, 1988). Moreover, since models of this type contain great technological detail, they can indicate that much can be done to significantly reduce energy consumption through a wide range of technological possibilities. Thus, they can also be used to look at supply- and demand-oriented policies aimed at curbing energy consumption and hence CO$_2$ emissions (cf. Beaver, 1993).

In modelling the implementation of available technologies, however, energy-sector optimization models consider the possibility of substitution among different options through absolute shifts. Therefore, they tend to underestimate transaction costs and to be too optimistic about potential for market penetration (Carraro *et al.*, 1994).[6] Moreover, although models of this type are able to show how the various exogenously-determined energy demands can be met at the least cost, the level of energy services demanded from the energy system is independent of prices. The lack of demand-price interactions is particularly troublesome in

---

[4] Not to be confused with dynamic programming which is used to solve in an optimal way complex sequential decision problems (Bellman, 1957). Here dynamic optimization models refer to those of an intertemporal structure, which maximize discounted utility or minimize discounted cost over the entire time horizon, subject to a set of constraints that tie together time-dependent variables.

[5] MARKAL is an acronym for MARKet ALlocation. For a detailed description of the model and some modifications, see, for example, Zhang (1992) and Chapter 8 of this book.

[6] At the consumer level, for example, market failures such as information costs and high discount rates can result in a limited adoption of available options.

models of this type if we anticipate that there will be considerable changes in relative prices caused by CO_2 emission limits (Manne and Wene, 1992). This also rules out the use of these models for an estimation of rebound effect.[7] In addition, because models of this type disregard intersectoral linkages (namely, there is no mutual interdependence of the energy sector and the rest of the economy), they do not say anything about the economic impacts of changes in relative prices caused, for example, by the introduction of a carbon tax. This would lead to incomplete and less reliable assessments and thus seriously flaw the analyses based on such models, in particular when analysing great changes brought about by a high carbon tax imposed in order to achieve a substantial cut in CO_2 emissions.

The second category refers to those optimization models with a detailed treatment of the energy sector but a highly aggregated description of the economy. These models are designed to remove some built-in limitations of the energy-sector optimization models and, at the same time, do not get lost in a high degree of technological detail. A good example is ETA-MACRO, which is also called GLOBAL 2100 in order to emphasize both the global nature of the carbon emission problem and the need for a long-term perspective (Manne and Richels, 1992). It is a merger between ETA (a process model for Energy Technology Assessment) and a MACROeconomic growth model with only one final output good in its highly aggregated representation of the economy. In recent modelling efforts, Manne and Wene (1992) replace ETA by MARKAL, which has considerably more technological details than ETA, and link MARKAL and MACRO. Thanks to the simplified representation of the overall economy in MACRO,[8] this linkage has been made possible. The linked MARKAL-MACRO model has been used as an analytical tool for Annex V (Energy Options for Sustainable Development) of the Energy Technology Systems Analysis Programme of the International Energy Agency (IEA-ETSAP) (cf. Kram, 1994).

In either GLOBAL 2100 or MARKAL-MACRO, the energy sector is linked to the rest of the economy in terms of an aggregated nested CES (constant elasticity of substitution) production function, with capital, labour, electric energy and non-electric energy as the four inputs. Thus, energy-economy interactions occur via

---

[7] The introduction of energy saving technologies does not necessarily lead to a proportionate decrease in energy demand because of substitution and income effects that partly offset the savings. The rebound effect, which is defined as the ratio of lost energy savings to theoretically expected energy savings, is just a means of quantifying this effect. See Jones (1993) and Kram (1994).

[8] The treatment makes it less difficult to gather a consistent international data set, while arriving at a meaningful summary of the results. It also lends itself to interpretation of the results, particularly for long-term projections and economic analysis of energy policies over a century or more on a global scale (Hogan and Jorgenson, 1991).

energy inputs into the economy in the production function and inter-industrial payments for energy costs. This makes GLOBAL 2100 and MARKAL-MACRO able to capture price-induced energy conservation and substitution among factors of production for the economy as a whole, thus satisfying the macroeconomic policy analysis. In addition, in the latest version of GLOBAL 2100, which is the applied general equilibrium extension of GLOBAL 2100, called 12RT (12-Region international Trade), the income flows and energy supply reactions between regions are modelled consistently. Thus the latest version can be used to analyse the feedback effects of international trade (cf. Manne, 1994).[9]

As an illustration of the present type of models the MARKAL model and the MARKAL-MACRO model have been compared with respect to a reduction of CO_2 emissions by 50% in 2030 relative to the 1990 level in the Netherlands. Figure 4.1 illustrates how the various steps contribute to reducing CO_2 emissions (Kram, 1994). The first two steps in MARKAL-MACRO clearly show a reduction in CO_2 emissions by price-induced cuts in GDP growth and specific useful energy demand, which cannot be estimated by MARKAL stand-alone. Moreover, the feedback effect of rising energy prices on the specific useful energy demand is subject to the MACRO elasticity of substitution (ESUB) parameter, with a higher ESUB yielding larger decreases in the amount of energy services demanded in MARKAL and hence in CO_2 emissions.

Although GLOBAL 2100 or MARKAL-MACRO allows for energy-economy interactions, its highly aggregated description of the economy means that neither GLOBAL 2100 nor MARKAL-MACRO can provide detailed information on the impacts of compliance with CO_2 emission limits on individual industries. This might be sufficient for global studies, but is clearly insufficient for a single-country study. Thus, this would to some extent limit its value as an evaluation tool for CO_2 constrained policy analysis.

4.4   Input-output and macroeconomic approaches

In this section, I shall discuss two classes of traditional economic models commonly used for the analysis of CO_2 emission limits: a) input-output (I-O) models, and b) macroeconomic models.

I start with the I-O models. The traditional I-O models describe systematically the complex sectoral interrelationships in an economy and record the many transactions taking place between the producing sectors of an economy by means of a set of easily solvable simultaneous linear equations. To be useful for environmental policy analysis, the traditional I-O models have been extended to

---

[9] The earlier version of GLOBAL 2100 allows also for international trade in carbon emission rights (Manne and Richels, 1992). However, the model suffers from lack of world trade consistency, because its regional submodels have not been linked.

take account of relationships between economic activities and the environment (see for example Miller and Blair (1985) and Pearson (1989) for a further discussion). Less ambitious extensions along this line have involved adding extra rows to represent the generation of pollutants, and sometimes extra columns to represent pollution abatement activities (cf. Leontief, 1970; Miller and Blair, 1985; Pearson, 1989).

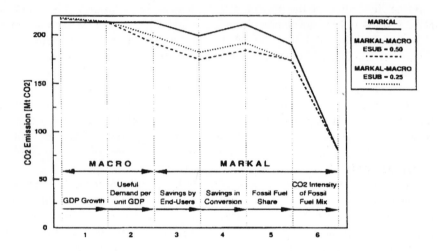

Source: Kram (1994).

Figure 4.1   *MARKAL versus MARKAL-MACRO: Steps to reduce CO$_2$ emissions*

In this way, I-O models have been used in a variety of applications, including the analysis of CO$_2$ emission limits. A good example is the work of Proops *et al.* (1993). It estimates the structural adjustments necessary to achieve a 20% reduction in CO$_2$ emissions over 20 years (approximately a 1% per annum reduction in CO$_2$ emissions) for Germany and the UK, using the input-output models for the German and UK economies of 47 sectors. Table 4.1 shows the results of this calculation when there are no GDP growth or employment growth constraints. The results suggest that to achieve the CO$_2$ emissions limits would come at the expense of about half that rate of reduction in GDP in both countries. With respect to the sectoral impact, it is also shown that all sectors are required to reduce final demand, with Electricity Distribution (Sector 5) recording the largest rate of reduction for both Germany (-2.56% p.a.) and the UK (-2.47% p.a.).

Table 4.1   *Changes in final demand (% change p.a.; -: declines)*

Sector	Germany	UK
1 Agriculture	-0.17	-0.09
2 Forestry and fishing	-0.03	-0.01
3 Electricity: fossil generation	0.00	0.00
4 Electricity: other generation	0.00	0.00
5 Electricity: distribution	-2.56	-2.47
6 Gas	-0.02	-0.05
7 Water	-0.01	-0.03
8 Coal extraction, coke ovens, etc.	-0.17	-0.05
9 Extraction of metalliferous ores	-0.01	0.00
10 Extraction of mineral oil and gas	0.00	-0.04
11 Chemical products	-1.19	-0.64
12 Mineral oil processing	-0.11	-0.24
13 Processing of plastics	-0.11	-0.04
14 Rubber products	-0.04	-0.03
15 Stone, clay, cement	-0.11	-0.05
16 Glass, ceramic goods	-0.10	-0.04
17 Iron and steel, steel products	-0.60	-0.20
18 Non-ferrous metals	-0.13	-0.09
19 Foundries	-0.02	-0.01
20 Production of steel, etc.	-0.09	-0.13
21 Mechanical engineering	-0.51	-0.23
22 Office machines	-0.07	-0.02
23 Motor vehicles	-0.77	-0.25
24 Shipbuilding	-0.02	-0.04
25 Aerospace equipment	-0.03	-0.08
26 Electrical engineering	-0.35	-0.20
27 Instrument engineering	-0.05	-0.02
28 Engineers' small tools	-0.17	-0.02
29 Music instruments, toys, etc.	-0.03	-0.02
30 Timber processing	-0.02	-0.02
31 Wooden furniture	-0.14	-0.07
32 Pulp, paper, board	-0.14	-0.03
33 Paper and board products	-0.06	-0.03
34 Printing and publishing	-0.02	-0.04
35 Leather, leather goods, footwear	-0.02	-0.02
36 Textile goods	-0.16	-0.08
37 Clothes	-0.08	-0.03
38 Food	-0.92	-0.48
39 Drink	-0.12	-0.14
40 Tobacco	-0.02	-0.02
41 Construction	-0.85	-0.67
42 Trade wholesale and retail	-0.83	-0.84
43 Traffic and transport services	-0.63	-0.64
44 Telecommunications	-0.05	-0.05
45 Banking, finance, insurance, etc.	-0.39	-0.22
46 Hotels, catering, etc.	-0.28	-0.21
47 Other services	-0.51	-0.07
GDP	-0.57	-0.43

Source: Proops *et al.* (1993).

I-O models contain a far higher degree of sectoral detail than the dynamic optimization models, macroeconomic models and computable general equilibrium models discussed below, because the computational capabilities of powerful PCs make it possible to solve I-O models with several dozens of sectors without complication and within reasonable time. Consequently, I-O models are most often used where it is important to analyse the detailed sectoral consequences of carbon abatement policies (cf. Fankhauser and McCoy, 1995). However, the high sectoral disaggregation of I-O models has its price. To be able to cope with a large number of sectors, I-O models impose a set of strong restrictions, including fixed input-output coefficients,[10] constant returns to scale, perfect factor supply, and exogenously-determined final demand. This, in turn, restricts I-O models to short-run analysis and means that I-O models paint a somewhat distorted picture of an essentially nonlinear world (Pearson, 1989; Fankhauser and McCoy, 1995).

Now I consider the macroeconomic models. Just as I-O models, macroeconomic models are demand-driven, but macroeconomic models go beyond I-O models by carefully modelling the role of prices and by incorporating the supply-side equilibrating mechanisms. Models of this type are neo-Keynesian in spirit in that the final demand remains the principal determinant of the size of the economy and in that macroeconomic models include neoclassical representation of the supply side of the economy. In macroeconomic models, the equilibrating mechanisms work through quantity adjustments rather than price adjustments as in the CGE models. Thus, macroeconomic models allow for temporary disequilibria in the markets for products, labour and foreign exchange, which are represented by the underutilization of production capacities, unemployment, and the imbalance on current account respectively (cf. Capros and Karadeloglou *et al.*, 1990a).

There are many macroeconomic models in existence. A good example is the HERMES model. It was constructed on behalf of the CEC initially for a Harmonized European Research for Macrosectoral and Energy Systems (cf. Italianer, 1986). HERMES has been expanded to accommodate issues surrounding emissions of pollutants, and is used to evaluate the economic consequences of the introduction of a carbon/energy tax.

---

[10] Under the assumption of fixed input-output coefficients, that is, the assumption of no substitution between various fuels and between fuels and other inputs in production, we expect that the carbon tax required to achieve the same reduction in CO_2 emissions is higher than if such a substitution is allowed. This has been confirmed by the study of Symons *et al.* (1994), the results of which suggest that within an input-output framework the carbon tax required to reduce 20% in CO_2 emissions attributable to household consumption of goods in the UK is higher than the estimates for carbon taxes that would achieve the Toronto target and where such substitution is allowed.

Table 4.2 shows the impact of the CEC tax[11] on GDP, unemployment and inflation for the four greatest economies of the EC using the HERMES model, separately and together. The three columns present scenarios under which the tax revenues are retained by the government to reduce public deficit or are recycled into the economy by means of reducing either personal income taxes or employers' social security contributions. It can be seen that the effects of recycling the tax revenues on both GDP and employment are markedly less negative than in the tax retention case and that the macroeconomic effects of reducing social security taxes are more positive than in the direct tax offset case. This highlights that the 'double dividend' feature of a carbon tax has important implications for 'green tax swaps' for distortionary taxes.[12]

Table 4.2   *Main macroeconomic effects of the CEC tax in 2005*

	Direct taxes	**Social security**	**Public deficit**
1. GDP (percentage deviations relative to baseline)			
France	-0.7	-0.3	-0.92
FR Germany	-0.6	-0.2	-1.37
Italy	-0.0	0.2	-2.19
United Kingdom	-0.7	-0.2	-2.05
Europe 4	-0.53	-0.12	
2. Unemployment rate (percentage deviations relative to baseline)			
France	-0.0	-0.2	0.03
FR Germany	-0.3	-0.7	-0.7
Italy	-0.0	-0.2	0.43
United Kingdom	0.4	-0.1	0.78
Europe 4	0.01	-0.37	
3. Consumer price index (percentage deviations relative to baseline)			
France	4.3	3.2	
FR Germany	2.7	2.0	
Italy	1.8	-0.5	
United Kingdom	5.5	3.1	
Europe 4	3.54	1.92	

Sources: Karadeloglou (1992) and Standaert (1992).

---

[11] For a description of the CEC tax, see Section 2.5.

[12] For a further discussion, see the treatment of the carbon tax revenues in Sections 2.5 and 7.4.

It is generally thought that macroeconomic models are often more related to reality than CGE models, because a number of interrelated equations in macroeconomic models are determined by the traditional econometric approach, which relies largely on reliable time-series data for sufficiently long periods (cf. de Melo, 1988; Boero *et al.*, 1991). Consequently, these models are able to capture the transitional impacts of exogenous shocks caused, for example, by the introduction of a carbon tax on such main economic indexes as inflation and unemployment, which are crucial and dominate the decision-making process in particular in the short run (cf. Borges, 1986). This makes macroeconomic models' a persuasive instrument in influencing decision making (Fankhauser and McCoy, 1995).

Behavioural patterns used in macroeconomic models, however, may be subject to what has been called the Lucas critique (Lucas, 1976) - past behaviour may no longer hold true in the light of rational expectations regarding policy actions.[13] If this is so, then the estimated effects of policy actions will be incorrect. Clearly, this critique indicates that macroeconomic models are an inappropriate tool for analysing the economic effects of great changes in the demand and/or supply structure of an economy and those questions of long-run nature. This weakness limits their applicability to the analysis of a high carbon tax, because the short-run focus of macroeconomic models does not coincide with the time required for a carbon tax to materialize and because their usefulness in analysing the effects of 'low' carbon taxes by no means guarantees their reliability in the analysis of high carbon taxes required for achieving a substantial cut in CO_2 emissions, say, the Toronto target.

4.5   Computable general equilibrium approach

An environmental policy aimed at curbing CO_2 emissions by means of a carbon tax will change the relative prices of goods. Carbon-free or low-carbon containing goods and services become cheaper than those of high carbon intensity. Such changes in the relative prices will lead to a shift away from high-carbon energy, away from energy towards capital and labour, and away from carbon intensive goods and services. This will have feedback effects on the economic structure and products mix, economic growth, the allocation of resources, and the distribution of income. Clearly, analysing such economy-wide impacts cannot be carried out within a partial equilibrium framework. Moreover, if the carbon taxes are used to achieve, for example, the Toronto target, they must be non-marginal and therefore cannot be estimated reliably by partial equilibrium approaches, I-O models or macroeconomic models, which can at best indicate the effects of 'low' carbon taxes. Thus, CGE models are called for.

---

[13] See also Mankiw (1990) for a good discussion of the Lucas critique in the context of the breakdown of the consensus in macroeconomics.

Indeed, from a theoretical point of view CGE models are preferred to I-O models and macroeconomic models, because CGE models are based on a solid microeconomic foundation. In CGE models, the behaviour of economic agents is modelled explicitly and is based on microeconomic optimization principles, whereas macroeconomic models pay less attention to economic theory and more attention to time-series data. CGE models often operate by simulating the operating of markets for factors, products and foreign exchange, with equations specifying supply and demand behaviour across all markets, and are endogenously solved for a set of equilibrium wages, prices and an exchange rate to clear these markets. CGE models are Walrasian in spirit in that the equilibrating mechanisms work through changes in relative prices. The equilibrium solution to CGE models produces a wealth of detailed information, including market clearing prices and quantities for sectoral output, investment, employment, foreign trade, energy consumption and CO_2 emissions.

While macroeconomic models pay specific attention to modelling transitional adjustment costs associated with policy changes or exogenous shocks, CGE models place emphasis on examining the economy in different states of equilibrium instead. Recognizing that it can take considerable time for prices to adjust to bring supply and demand back into equilibrium, CGE models are essentially long-run in conception (Boero *et al.*, 1991; Fankhauser and McCoy, 1995). Moreover, in CGE models, parameters in utility functions and production functions are structural parameters, representing tastes and technologies. This makes CGE models less vulnerable to the Lucas critique (Bovenberg, 1985). Further, CGE models include a government sector, so the effects of alternative means of recycling the revenues generated by carbon taxes can be analysed.[14] For example, Jorgenson and Wilcoxen (1993a) have used an intertemporal CGE model of the United States, which has a more solid econometric foundation than most other CGE models and pays close attention to price-response technological change in different industries, to examine the costs of carbon taxes under three specifications for revenue recycling. In the first specification, the carbon tax revenues are recycled as lump-sum payments to households; in the second they are used to lower taxes on labour; and in the third the revenues are used to lower taxes on capital. The results of the three simulations show that a 1.7% GNP loss under lump-sum redistribution is converted to a 0.7% loss in the second simulation and a 1.1% gain in the third one. The former is due to an increase in employment brought about by the drop in the wedge between before- and after-tax wages, whereas the latter is due to accelerated capital formation generated by an increase in the after-tax rate of return on investment. Another good example of using CGE models to examine interactions between carbon taxes and pre-existing taxes on labour and capital is the work of Goulder (1995). The Goulder

---

[14] See Section 2.5.3 for a discussion about alternative carbon tax recycling options and Section 7.4 for analysing the economic impact of carbon taxes for China under the four indirect tax offset scenarios.

model, though an intertemporal CGE model of the United States, differs from the Jorgenson and Wilcoxen model, not least because it contains considerably more detail on US taxes. The salient feature makes it especially well suited to analysing the costs of carbon taxes in the presence of prior tax distortions. The Goulder study indicates that incorporating preexisting taxes on labour and capital results in the higher costs of carbon taxes than would be the case if there were no preexisting factor taxes. This is because a carbon tax operates much like an implicit factor tax and thus increases further distortions in factor markets (cf. Bovenberg and de Mooij, 1994). Clearly, this effect serves to blunt the effectiveness of revenue recycling as a means of lowering the net costs of carbon taxes.

From the preceding discussion, it is not surprising that CGE models are widely used to analyse the economic effects of limiting CO_2 emissions, at both national and international levels. Examples of national level are the works of Bergman (1991) for Sweden, Ingham and Ulph (1991) for the United Kingdom, Glomsrød et al. (1992) for Norway, Proost and van Regemorter (1992) for Belgium, Stephan et al. (1992) for Switzerland, Conrad and Schröder (1991, 1993) for Germany, Jorgenson and Wilcoxen (1993a, 1993b) and Goulder (1995) for the United States, and Beauséjour et al. (1995) for Canada, while global studies based on CGE models are exemplified by the works of Whalley and Wigle (1991a, 1991b), Burniaux et al. (1992), Martin et al. (1992), Martins et al. (1992, 1993), Pezzey (1992), and Piggott et al. (1992, 1993).

For analysing the economy-wide impacts of the predefined Chinese energy-related CO_2 emission limits, I have chosen the CGE approach. This choice has been motivated by the wide recognition of CGE approach as an appropriate tool for such a purpose. In Chapter 5 of this book, first a CGE model of the Chinese economy is described, followed by some work done for empirical application of the CGE model in Chapter 6. Its application in macroeconomic analysis of CO_2 emission limits for China is presented in Chapter 7.

CGE models have, of course, some limitations for practical policy decisions. The most frequently mentioned one is the lack of empirical validation (Borges, 1986). Although there are some exceptions, most CGE models are calibrated rather than econometrically estimated.[15] The calibration procedure often borrows a variety of elasticities from other studies (see Section 6.4 for a further discussion). Whatever values of elasticities are chosen, they are difficult to defend, because often these studies do not contain the same definitions of variables or level of disaggregation. The second weakness concerns the general equilibrium assumption. This assumption, combined with that CGE models stress relative

---

[15] Among the examples adopting the traditional econometric approach to CGE modelling are the DGEM model for the United States (Jorgenson and Wilcoxen, 1993a, 1993b), and the work of Glomsrød et al. (1992) for Norway and McKitrick (1995) for Canada.

prices,[16] rules out the use of CGE models for analysing traditional disequilibrium issues such as inflation and unemployment (Borges, 1986; Boero *et al.*, 1991). Another limitation built into most CGE models is the assumption of perfect competition.[17] Clearly, this is not representative of the real world, where many cases of market failure exist, such as monopoly power and imperfect competition.

4.6   Hybrid approach

As discussed earlier, bottom-up models[18] such as energy-sector optimization models take a disaggregate approach to modelling energy supply and demand (Wilson and Swisher, 1993). They can identify, for example, the potentials of energy efficiency improvement to which each energy technology will contribute, and provide information on the corresponding costs required to achieve such potentials. In bottom-up models, however, the transaction costs associated with implementing technologies are underestimated and the feedbacks from and to other sectors of an economy are not included. By contrast, top-down models such as macroeconomic models and CGE models take a macroeconomic approach to modelling energy-economy interactions and the costs of changing them (Wilson and Swisher, 1993). They can provide detailed information on the impacts on individual industries. But in top-down models, future energy demand and the cost

---

[16] All supply and demand functions in CGE models that are mainly used for the allocation of resources are assumed to be homogeneous of degree zero in prices. As a consequence, only relative prices are important for the determination of the quantities of goods supplied and demanded. Thus, CGE modellers usually choose a price index as the price numéraire, and all other prices are measured relative to it. For a further discussion, see Section 5.3.9.

[17] In principle, this weakness can be overcome. To cite an example, Harris (1984) introduces economies of scale in production and imperfect competition in an applied general equilibrium model for the analysis of trade liberalization. This example remains an exception, however, in the sense that it has not had many followers among CGE modellers. The reason would be that incorporating economies of scale and imperfect competition into CGE models, though useful for a policy study based on a CGE model, would complicate the analysis. Thus, for simplicity, a number of CGE modellers adopt the explicit assumption of constant returns to scale in production and perfect competition.

[18] The so-called bottom-up models refer to those models that typically incorporate a detailed representation of technologies for energy supply and use, but little representation of markets and none of the rest of the economy. Thus these models are often referred to as the engineering models.

of changing it are to a large extent determined by two macroeconomic parameters, which are respectively known as autonomous (that is, non-price-induced) energy efficiency improvement and the elasticity of price-induced substitution between the inputs of capital, labour and energy[19] and which are neither physically observable nor directly measurable (Wilson and Swisher, 1993). Consequently, it is impossible for top-down models to indicate from which energy conservation originates and to convince energy planners how it can be achieved.

Given the relative strengths and weaknesses of bottom-up models and top-down models, these two approaches, rather than competing or substituting, can certainly serve to complement each other if they are linked. Because of operational difficulties in directly linking models of the two types, less ambitious attempts in this direction have involved informal linkage between the existing bottom-up models and top-down models. The purpose is to establish a consistent interaction between these two models, thus shedding light on both economic and technological aspects of the control of CO_2 emissions. Clearly, results based on this approach can satisfy both environmental policy analysis and energy planning requirement.

In informally-linked systems, these two models are operated as parallel independent units, but the results from one model run can be reflected in the other model to arrive at consistent scenarios for economic development, fuel choice, cost-efficient mix of energy technologies, and CO_2 emissions. The implementation following this approach has been exemplified by an early study, which integrates MENSA, an Australian regionalized version of MARKAL, with an Australian input-output model MERG (James *et al.*, 1986). In the informally-linked MENSA-MERG system, MENSA aims to define the most efficient structure for the allocation of fuels to the various end-use sectors. The evolution of structural shifts in the energy system exhibited in the MENSA solution would then be reflected in the MERG model as time-dependent intersectoral energy coefficients. These in turn would modify the end-use demands for energy within MENSA. Thus, the calculations would proceed iteratively until some convergence criterion was satisfied. Recent examples in this tradition are the ongoing works of various IEA-ETSAP member countries, which are performed to link a member country version of MARKAL with a macroeconomic model of its own (cf. Kram, 1993b). In this way the energy and environmental effects can be addressed in relation to the macroeconomic effects. In Chapter 8 of this book, a CGE model of the Chinese economy is linked with a technology-oriented model for power system expansion planning. Such a linkage makes it possible to study how China's electricity sector is able to comply with the CO_2 emission limits and the resulting macroeconomic implications.

---

[19] For a further discussion of two macroeconomic parameters, see Section 5.8.1.

If a bottom-up model is a simplified energy model, a direct linkage is possible.[20] A good example is the linked HERMES-MIDAS model (Capros *et al.*, 1990b). This linkage occurs by eliminating energy equations from HERMES and macroeconomic equations from MIDAS (Multinational Integrated Demand And Supply), the latter being a country-specific energy demand and supply model developed for European countries under the auspices of the CEC. Unlike the linkage between MENSA and MERG mentioned above, the linked HERMES-MIDAS model is considered a single model, since the linkage has been constructed to be formal and numeric by means of interface modules performing transformations of linked variables. This linked model has been used to evaluate the economic consequences of the introduction of a carbon or energy tax for both the economic and energy systems for the four greatest economies of the EC (cf. Karadeloglou, 1992).

While a hybrid approach is able to shed light on both economic and technological aspects of the control of CO_2 emissions, it does present some drawbacks. In order to obtain consistent linking results, a hybrid approach needs to remove all the inconsistencies built into the two models. This often turns out to be cumbersome and time-consuming. Moreover, top-down models are very different from bottom-up models in terms of the discipline from which they originate, which reflects that each modeller approaches the cost estimates for the control of CO_2 emissions starting with the best understood aspects. Thus, a hybrid approach adds an extra requirement for cooperation between modellers from different disciplines.

4.7   Concluding remarks

From the preceding analysis, the following main conclusions can be drawn. First, if focus is primarily placed on technological solutions to CO_2 emission problems, dynamic optimization models are very useful. Moreover, in order to prioritize investments in carbon abatement technologies, specific cost-effective analysis of these technologies is helpful. In this respect the *ad hoc* approach may be used. Second, of a variety of models discussed in this chapter, none contains more sectoral detail than input-output models. Therefore, if interest centres mainly on the consequences of a carbon tax for the economic structure, input-output models are generally considered an appropriate tool for such a purpose. Third, the transitional impacts of a carbon tax on inflation and unemployment can best be captured in macroeconomic models. Thus, if focus is placed on an estimation of transitional adjustment costs in the short run, we can rely on macroeconomic models. Fourth, CGE models are an appropriate tool for analysing the economic

---

[20] As discussed in Section 4.3, directly linking a simplified top-down model with a complicated bottom-up model has been exemplified by the MACRO-MARKAL model.

effects of great changes in the demand and/or supply structure of an economy and those questions of long-run nature. If we want to shed light on long-run aspects of a high tax imposed for achieving a substantial cut in CO_2 emissions, CGE models are called for. It is for this reason that a CGE approach has been chosen for analysing the macroeconomic impacts of CO_2 emission limits for China, which is the subject of Chapters 5 to 7. Finally, given the relative strengths and weaknesses of bottom-up models and top-down models, it is worthwhile to link them together so that they complement each other, thus shedding light on both economic and technological aspects of the control of CO_2 emissions. This has motivated me to link a CGE model of the Chinese economy with a technology-oriented model for power system expansion planning in China's electricity sector, which is discussed in Chapter 8.

5 A Computable General Equilibrium Model for Energy and Environmental Policy Analysis[1]

5.1 Introduction

This chapter describes a computable general equilibrium (CGE) model for energy and environmental policy analysis of the Chinese economy. There were at least three reasons for choosing a CGE approach to carry out this case study for China. First, the transition in China is taking place from a centrally planned economy to a market-oriented one, the latter relying on the increased use of market mechanisms and price incentives. The transitional character of the economy needs a proper treatment of the agents or actors in the economy, thus calling for different structural models given the fact that the traditional modelling approaches, such as input-output models and Keynesian demand-driven macroeconomic models, are not the appropriate tools for analysing economy-wide effects of great changes in the demand and/or supply structure and for those questions of long-term nature. The second reason is related to the availability of data. Standard econometric models require reliable time-series data for sufficiently long periods (cf. de Melo, 1988). The data are neither available nor appropriate for standard econometric analysis without considerable further preparation to remove inconsistencies. Third, in analysing the economic impacts of limiting CO_2 emissions, it has been argued in Chapter 4 that a CGE approach is generally considered an appropriate tool.

The CGE model of the Chinese economy operates by simulating the operation of markets for factors, products and foreign exchange. It is highly non-linear, with equations specifying supply and demand behaviour across all markets. Moreover, with focus being placed on addressing such energy and environmental issues as quantifying the economy-wide effects of policies aimed at limiting CO_2 emissions, our model pays particular attention to modelling the energy sector and its linkages to the rest of the economy, because the CO_2 emissions from fossil fuel combustion in the energy system are the main source of man-made CO_2 emissions, which in turn are the major cause of the greenhouse effect. This makes our CGE model different from other CGE models for China in several

---

[1] This chapter forms the inputs to two papers in *Journal of Policy Modeling* (Zhang, 1997b) and in *Economic Systems Research* (Zhang, 1997c).

99

aspects.[2] In our CGE model, energy use is disaggregated into coal, oil, natural gas and electricity. Along with capital, labour and intermediate inputs, the four energy inputs are regarded as the basic inputs into the nested constant elasticity of substitution-Leontief production function. Moreover, our model incorporates an explicit time dimension, and has a transparent representation of the rate of autonomous energy efficiency improvement (AEEI) unrelated to energy price increases if dynamic linkages proceed. So, the effect of the AEEI parameter can easily be assessed. Thus, our CGE model, which is also rich in treatment of foreign trade and is appropriate for modelling the household consumption, allows endogenous substitution among energy inputs and alternative allocation of resources as well as endogenous determination of foreign trade and household consumption in the Chinese economy in order to cope with the environmental restrictions, at both sectoral and macroeconomic levels. The equilibrium solution to the model for a given year produces a wealth of detailed information, including market clearing prices, GNP, productivity levels by industry, investment by industry, final consumption levels by commodity, employment by industry, imports and exports by commodity, fuel-specific production in physical terms, energy consumption patterns, and CO_2 emissions. Moreover, the Hicksian equivalent variation is calculated to measure the welfare impacts of, say, emission abatement policies. Furthermore, the CGE model incorporates an explicit tax system. This makes it suitable for estimating the 'double dividend' from the imposition of a carbon tax.

The remainder of this chapter proceeds as follows. Sections 5.2 to 5.10 describe the model equations for the following blocks: production and factors, prices, income, expenditures, investment and capital accumulation, foreign trade, energy and environment, welfare measures, and market clearing conditions and macroeconomic balances. In Section 5.11 some concluding remarks with respect to directions for further work are drawn. In describing the equations, the endogenous variables are denoted by capital letters, whereas the exogenous ones or

---

[2] The literature on the development of CGE models for China is growing. Byrd (1989) examined the impacts of planning and the efficiency of markets in the two-tier plan/market system in Chinese state-owned industry using a static general equilibrium model. Because it is purely analytic, Byrd's model is unable to provide any indication of the order-of-magnitude of the effects of policy changes. Martin (1990) used a linearized CGE model of the ORANI type (Dixon *et al.*, 1982) to analyse the effects of exchange rate devaluation on the Chinese economy. Revising the data base of 1981 from the World Bank (1985b), Shi (1991) constructed a 8-sector static CGE model for energy and foreign trade policy analysis. The recent CGE model of the Chinese economy is that of Garbaccio (1994), which includes an explicit representation of the two-tier plan/market system and is used to analyse the sectoral effects of reforms in pricing and taxation. All these models lack dynamics and are used for comparative-static analyses.

parameters are written in lower-case letters, Greek letters and letters with a bar. The indices *i* and *j* refer to sectors or goods, *t* in parentheses to time period, and *h* to type of household. The meaning of a symbol is explained on first use.

5.2 Production and factors

The production block includes ten sectors: agriculture, heavy industry, light industry, transport, construction, services, coal, oil, natural gas and electricity. The first six sectors are associated with the production of goods and services, while the last four relate to the supply and distribution of energy. All sectors are assumed to operate at constant returns to scale.[3] In each sector gross output is produced using four energy inputs, capital, labour and intermediate goods and services, with the substitution taking place across energy inputs, capital and labour.

5.2.1 Production function

The technology of production is represented by nested constant elasticity of substitution (CES)-Leontief function. It combines Leontief specifications for non-energy intermediate goods with a nested CES aggregate of composite capital-labour input and composite energy input. The structure of the production function is the same for all sectors, but the elasticities of substitution between various inputs may differ across sectors.

The nesting hierarchy of our CGE model, based on separability assumptions, is depicted in Figure 5.1.[4] Starting from the bottom, the production function is characterized by Cobb-Douglas (CD) aggregations of capital and labour to form a value added aggregate (that is, composite capital-labour input), and intermediate inputs of coal, oil, natural gas and electricity to form an energy aggregate (that is, composite energy input). Then, in turn, these two aggregates combine, by means of a CES aggregation function, into the composite input $EVA_i(t)$. Finally,

---

[3] Harris (1984) introduces economies of scale in production in an applied general equilibrium model for the analysis of trade liberalization. This example remains an exception, however, in the sense that it has not had many followers among CGE modellers. The reason would be that incorporating economies of scale into CGE models, though useful for a policy study based on a CGE model, would complicate the analysis. Thus, for simplicity, CGE modellers adopt the explicit assumption of constant returns to scale in production.

[4] Nesting refers to functional specification that imposes separability between bundles of factors and goods. This makes it possible to separate optimizing decisions of agents into several stages.

this composite input is combined with non-energy intermediate inputs to produce gross output $Q_i(t)$.

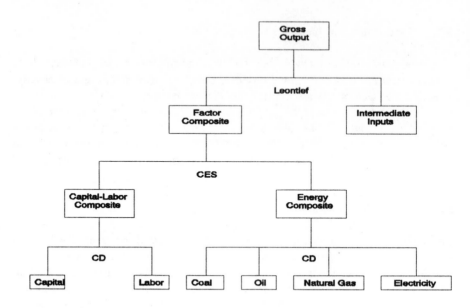

Figure 5.1   *Nesting structure of production in the CGE model*

Thus the nested CES-Leontief production function can be expressed as follows:

$$VA_i(t) = \overline{A}_i e^{\lambda_i(t)*(t-t_o)} K_i(t)^{\alpha_i} L_i(t)^{1-\alpha_i}$$

where

$VA_i(t)$	=	Composite capital-labour input of sector i in period t.
$\overline{A}_i$	=	Shift parameter associated with composite capital-labour input of sector i.
$K_i(t)$	=	Fixed capital stock of sector i in period t.
$L_i(t)$	=	Employment by sector i in period t.
α_i	=	Share of capital input in the value added aggregate of sector i.
t_o	=	Base year.
$\lambda_i(t)$	=	Productivity growth rate in sector i in period $(t-t_o)$.
e	=	Exponential function.

$$E_i(t) = \overline{B}_i \cdot \prod_{j=7}^{10} [e^{aei_i(t)*(t-t_o)} VE_{ji}(t)]^{b_{ji}}$$

where

$E_i(t)$	=	Composite energy input of sector i in period t.
$\overline{B}_i$	=	Shift parameter associated with composite energy input of sector i.
$VE_{ji}(t)$	=	Intermediate input of energy j by sector i in period t.
b_{ji}	=	Share of energy input of type j in the energy aggregate of sector i, with $\Sigma_{j=7}^{10} b_{ji}=1$.
$aei_i(t)$	=	Autonomous efficiency improvement in energy use of sector i in period $(t-t_o)$.

$$EVA_i(t) = \Omega_i \left\{ \omega_i VA_i(t)^{\rho_i} + (1-\omega_i)E_i(t)^{\rho_i} \right\}^{1/\rho_i}$$

where

$EVA_i(t)$	=	Aggregate of inputs from the composite capital-labour and the composite energy in sector i in period t.
Ω_i	=	Efficiency parameter associated with composite input of capital, labour and energy aggregate in sector i.
ω_i	=	Distribution parameter associated with value added aggregate in sector i.
ρ_i	=	$(\sigma_i-1)/\sigma_i$, where σ_i is the elasticity of substitution between the value added aggregate and the energy aggregate in sector i.

$$Q_i(t) = \sum_{j=1}^{6} a_{ji}Q_i(t) + EVA_i(t)$$

where

$Q_i(t)$	=	Gross output of sector i in period t.
a_{ji}	=	Input-output coefficients that represent intermediate requirements from sector j per unit of output of sector i.

5.2.2 Unit costs

The unit costs of the composite capital-labour input $VA_i(t)$ and the composite energy input $E_i(t)$ as well as the composite input $EVA_i(t)$ are the duals of production functions above:[5]

$$CVA_i(t) = \frac{1}{\bar{A}_i e^{\lambda_i(t)*(t-t_o)}} \left(\frac{UK_i(t)}{\alpha_i} \right)^{\alpha_i} \left(\frac{W_i(t)}{1-\alpha_i} \right)^{1-\alpha_i}$$

where
$CVA_i(t)$ =   Unit cost of composite capital-labour input of sector i in period t.
$UK_i(t)$ =   User cost of capital in sector i in period t.
$W_i(t)$ =   Wage rate in sector i in period t.

$$CE_i(t) = \frac{1}{\bar{B}_i} \prod_{j=7}^{10} \left(\frac{PF_j(t)}{e^{aei_i(t)(t-t_o)} b_{ji}} \right)^{b_{ji}}$$

where
$CE_i(t)$ =   Unit cost of composite energy input of sector i in period t.
$PF_j(t)$ =   User price of fuel of type j in period t.

---

[5] Suppose that the production function of the Cobb-Douglas form is given by

$$y = A \prod_i (b_i x_i)^{a_i},$$

where A denotes the shift parameter, b_i the augmentation parameter, a_i the share parameter, and x_i the input. Given the input prices w_i, the producers are assumed to minimize their production costs subject to the production function. Thus, the first-order conditions for cost minimization yield the following corresponding cost function $c(w_1, w_2, \ldots, w_n, y)$ and demand functions for input i:

$$c(w_1, w_2, \ldots, w_n, y) = \frac{1}{A} \prod_{i=1}^{n} \left(\frac{w_i}{a_i b_i} \right)^{a_i} y$$

$$x_i = \frac{a_i c(w_1, w_2, w_i, \ldots, w_n, y)}{w_i}$$

$$CEVA_i(t) = \Omega_i^{-1}\left(\omega_i^{\sigma_i}CVA_i(t)^{(1-\sigma_i)} + (1-\omega_i)^{\sigma_i}CE_i(t)^{(1-\sigma_i)}\right)^{1/(1-\sigma_i)}$$

where

$CEVA_i(t)$ =   Unit cost of the composite input $EVA_i(t)$ of sector i in period t.

5.2.3 Demands for primary factor and intermediate goods

Given the technology described above, producers are assumed to minimize their production costs. Thus, the first-order conditions for cost minimization yield

$$\frac{UK_i(t)}{W_i(t)} = \frac{\alpha_i}{1-\alpha_i} \frac{L_i(t)}{K_i(t)}$$

$$\frac{PF_r(t)}{PF_s(t)} = \frac{b_{ri}}{b_{si}} \frac{VE_{si}(t)}{VE_{ri}(t)}$$

$$\frac{CVA_i(t)}{CE_i(t)} = \frac{\omega_i}{1-\omega_i} \left(\frac{VA_i(t)}{E_i(t)}\right)^{-1/\sigma_i}$$

Substituting the first-order conditions above into production functions and using the unit cost definitions, it is possible to derive the following optimal demands for the inputs of labour, capital and intermediate energy of type j:

$$L_i(t) = (1-\alpha_i)PVA_i(t)Q_i(t)/W_i(t)$$

$$K_i(t) = \alpha_i PVA_i(t)Q_i(t)/UK_i(t)$$

$$E_i(t) = \left(\frac{1-\omega_i}{\omega_i}\right)^{\sigma_i} \left(\frac{CVA_i(t)}{CE_i(t)}\right)^{\sigma_i} VA_i(t)$$

$$VE_{ji}(t) = \frac{b_{ji}}{\alpha_i} \frac{1-\omega_i}{\omega_i} \left(\frac{E_i(t)}{VA_i(t)}\right)^{\rho_i} \frac{UK_i(t)K_i(t)}{PF_j(t)}$$

where

$PVA_i(t)$ =   Value added or net price of sector i in period t.

The intermediate goods are determined in each non-energy sector by fixed input-output coefficients, but in each energy sector by summing over the demands for intermediate energy of type i by sector j. Specifically,

$$V_i(t) = \sum_{j=1}^{10} a_{ij} Q_j(t) \qquad i = 1,...,6$$

$$V_i(t) = \sum_{j=1}^{10} VE_{ij}(t) \qquad i = 7,...,10$$

where

$V_i(t)$ =   Demands for intermediate goods by sector i in period t.

5.3 Prices

The price block presents all the price equations.

5.3.1 Prices of imports and exports

The domestic prices of imports are in the Chinese currency yuan and include *ad valorem* taxes, while the world market prices are in US dollars and are exogenously determined under the small-country assumption. The exchange rate, which is the price of a dollar in terms of the Chinese currency, is used to convert US dollars into yuan. Thus, the domestic prices of imports in a tariff-ridden economy are determined by

$$PM_i(t) = (1+tm_i) \cdot \overline{PWM}_i(t) \cdot ER(t)$$

where

$PM_i(t)$ =   Domestic price of imports of good i in period t.

tm_i =   Import tariff rate of good i.

$\overline{PWM}_i(t)$ =   World (dollar) price of imports of good i in period t.

$ER(t)$ =   Exchange rate between US\$ and domestic currency in period t.

Analogous to the import price relation, the domestic prices of exports can be written as:

$$PX_i(t) = (1+te_i) \cdot PWX_i(t) \cdot ER(t)$$

where

$PX_i(t)$ =   Domestic price of Chinese exports by sector of origin i in period t.

$PWX_i(t)$ =   World (dollar) price of Chinese exports by sector of origin i in period t.

te_i =   Export subsidy rate of good i.

5.3.2   Price of composite commodity

Under the assumption of cost minimization by the users of imports and domestic goods, the price of composite commodity is determined by the corresponding unit cost function that is dual to the CES aggregation function, namely:

$$P_i(t) = \Psi_i^{-1} \left(\mu_i^{\psi_i} PM_i(t)^{(1-\psi_i)} + (1-\mu_i)^{\psi_i} PD_i(t)^{(1-\psi_i)} \right)^{1/(1-\psi_i)}$$

where

$P_i(t)$	=	Composite price of commodity i in period t.
$PD_i(t)$	=	Price of domestic good i in period t.
Ψ_i	=	Shift parameter associated with composite commodity i in import demand function.
μ_i	=	Share parameter associated with imported good i in import demand function.
ψ_i	=	Price elasticity of substitution between imported and domestically-produced commodity i.

5.3.3   Domestic sale price

The sale price of domestically-produced commodity i is defined as an average of domestic price of exports $PX_i(t)$ and domestic good price $PD_i(t)$:

$$PS_i(t) = (PX_i(t)X_i(t) + PD_i(t)D_i(t))/Q_i(t)$$

where

$PS_i(t)$	=	Sale price of domestically-produced commodity i in period t.
$X_i(t)$	=	Total exports of commodity i in period t.
$D_i(t)$	=	Total domestic demand for domestic commodity i in period t.

5.3.4   Sectoral net price

The sectoral net price is defined as the output price minus indirect taxes and the cost of intermediate inputs:

$$PVA(t) = PS_i(t) \cdot (1 - itax_i) - \sum_j a_{ji} P_j(t)$$

where

$itax_i$	=	Indirect enterprise tax rate of sector i.

5.3.5 Price of capital services

The price of capital services in sector i is calculated as a weighted average according to the composition of capital used in that sector (by sector of origin):

$$PK_i(t) = \sum_j sf_{ji} \cdot P_j(t)$$

where
$PK_i(t)$ = Price of fixed capital goods in sector i in period t.
sf_{ji} = Fixed capital composition coefficients that represent the share of sector j in total fixed capital investment of sector i.

$PK_i(t)$ differs across sectors, reflecting the fact that capital used in different sectors is heterogeneous.

5.3.6 User price of capital

Once $PK_i(t)$ is determined, the user price of capital is calculated by

$$UK_i(t) = (\delta_i + R) \cdot PK_i(t)$$

where
δ_i = Depreciation rate of fixed assets in sector i.
R = Real rate of interest.

5.3.7 User price of fuels

The user prices of fuels are determined by the following formula:

$$PF_j(t) = (1 + \tau f_j(t)) \cdot P_j(t)$$

where $\tau f_j(t)$ is the *ad valorem* tax rate on fuel j in period t. It is converted from a given carbon tax. See the energy and environment block.

5.3.8 Price of consumer goods

Denoting the transformation matrix by tr_{ij} (see Section 5.5.1 for a further discussion of this matrix), the prices of consumer goods are determined by

$$PC_j(t) = \sum_{i=1}^{6} tr_{ij} \cdot P_i(t) + \sum_{i=7}^{10} tr_{ij} \cdot PF_i(t)$$

where
$PC_j(t)$ = Price of consumer good j in period t.

5.3.9 Price numéraire

All supply and demand functions in the CGE model are assumed to be homogeneous of degree zero in prices. As a consequence, only relative prices are important for the determination of the quantities of goods supplied and demanded. Moreover, we focus on multisectoral issues of growth, resource allocation, and structural change rather than analysing the causes of inflation. Thus, the GNP price deflator $\overline{PINDEX}(t)$ is chosen as the price numéraire.[6] It equals nominal GNP $Y(t)$ divided by real GNP $RY(t)$. Thus, the normalization equation will take the form

$$\frac{Y(t)}{RY(t)} = \overline{PINDEX}(t)$$

where

$\overline{PINDEX}(t)$	=	GNP price deflator in period t.
$Y(t)$	=	Total nominal GNP in period t.
$RY(t)$	=	Real GNP in period t.

$\overline{PINDEX}(t)$ is fixed exogenously and all other prices are measured relative to it. For simplicity, all prices are assumed to be equal to one in the base year. In this way, the benchmark data are set in value terms, with no need to specify the underlying volumes.

5.4   Income

The income block determines national income and the distribution of income among enterprises, households and the government.

5.4.1 National income

Nominal national income or nominal GNP is calculated at market prices. It is determined as the sum of value added plus indirect taxes and tariffs less export subsidies. Thus, we have

---

[6] The GNP deflator is a convenient choice. Any other price index could be used as the price numéraire. Other common choices in CGE models include a consumer or producer price index, exchange rate, or wage (cf. Dixon *et al.*, 1982; Dixon and Parmenter, 1994; de Melo and Robinson, 1989; Jorgenson and Wilcoxen, 1989).

$$Y(t) = \sum_i [PVA_i(t) \cdot Q_i(t)] + INDT(t) + TARIFF(t) - NETSUB(t)$$

where

$INDT(t)$	$=$	Total indirect tax revenue in period t.
$TARIFF(t)$	$=$	Tariff revenue in period t.
$NETSUB(t)$	$=$	Total export subsidies in period t.

By contrast, real national income or real GNP is calculated from the expenditure side. It is defined as

$$RY(t) = \sum_i \left(\sum_h CI_{hi}(t) + \overline{G}_i(t) + FI_i(t) + SK_i(t) + (1 - te_i) \cdot X_i(t) - (1 - tm_i) \cdot M_i(t) \right)$$

where

$CI_{hi}(t)$	$=$	Final demand for goods from sector i by household h in period t.
$\overline{G}_i(t)$	$=$	Government consumption of goods supplied by sector i in period t.
$FI_i(t)$	$=$	Volume of fixed investment goods supplied by sector i in period t.
$SK_i(t)$	$=$	Circulating capital investment by sector i in period t.
$M_i(t)$	$=$	Total imports in sector i in period t.

In the base year, $Y(t) = RY(t)$. Thus, the model calibration to be discussed in the next chapter should make sure that this condition is satisfied.

5.4.2 Factor income

The factor income includes labour income and capital income. The labour income is simply the product of the wage rate and employment:

$$WB_i(t) = W_i(t) \cdot L_i(t)$$

where

$WB_i(t)$	$=$	Wage bill or labour income in sector i in period t.

The wage rates in each sector are assumed to grow at the same rate. This specification for structural rigidity of sectoral wage rates reflects the wage match behaviour among sectors (cf. Chen, 1990). Moreover, the overall growth rate of wage in each period is assumed to be flexible enough to clear the labour market which is treated as though it is competitive. In nominal terms, the wage rates thus adjust according to the Laspeyres price index of consumer goods:

$$W_i(t) = (1 + GRW(t)) \cdot wo_i \cdot CPI(t)$$

where
GRW(t) = Overall growth rate of wage in period t.
wo_i = Base year wage rate in sector i.
CPI(t) = Laspeyres consumer price index in period t, defined as

$$CPI(t) = \frac{\sum_i PC_i(t) \cdot co_i}{\sum_i pco_i \cdot co_i}$$

where
co_i = Base year consumption of consumer good i.
pco_i = Base year price of consumer good i.

Capital income is determined as the difference between the value added minus labour income, namely

$$YK_i(t) = PVA_i(t) \cdot Q_i(t) - WB_i(t)$$

where
$YK_i(t)$ = Capital income in sector i in period t.

5.4.3 Net enterprise income

Net enterprise income or distributed profits to households is defined as capital income and transfer payments from the government, net of enterprise taxes, retained earnings which go to the capital account, and depreciation (Robinson *et al.*, 1990).

$$YE(t) = \sum_i YK_i(t) + \overline{GENT}(t) - ESAV(t) - ETAX(t) - DEPR(t)$$

where
$YE(t)$ = Net enterprise income in period t.
$\overline{GENT}(t)$ = Government transfer payments to enterprises in period t.
$DEPR(t)$ = Total value of depreciation of fixed capital stock in period t.
$ETAX(t)$ = Enterprise tax revenue in period t.
$ESAV(t)$ = Total nominal enterprise savings in period t.

$ESAV(t)$ is in turn determined through a fixed rate of enterprise retained profits:

$$ESAV(t) = esr(t)\left(\sum_i YK_i(t) + \overline{GENT(t)} - ETAX(t) - DEPR(t)\right)$$

where

$esr(t)$ =   Enterprise saving rate in period t.

5.4.4 Household income

The income received by households includes labour income, distributed profits from enterprises, transfers from the government, and remittances from abroad. In the context of CO_2 emissions, a part of the carbon tax revenues may be redistributed to households in order to compensate them from suffering from the carbon tax. By apportioning income from each source to the households of different types according to the fixed shares, total income received by the different household categories in period t can thus be written as

$$Y_h(t) = ws_h(t)\sum_i WB_i(t) + es_h(t)\cdot YE(t) + ts_h(t)\cdot\overline{HHT(t)} + rs_h(t)\cdot ER(t)\cdot\overline{REMIT(t)}$$

$$+ rt(t)\cdot RC(t)\left\{\overline{POP}_h(t)/\sum_h \overline{POP}_h(t)\right\}$$

where

$Y_h(t)$	=	Total income of household h in period t.
$\overline{REMIT(t)}$	=	Net remittances from abroad in US dollars in period t.
$\overline{HHT(t)}$	=	Government transfer payments to households in period t.
$RC(t)$	=	Total government revenue from the proceeds of tax levied on carbon emissions in period t.
$es_h(t)$	=	Share of household h in total distributed profits from enterprises in period t.
$rs_h(t)$	=	Share of household h in total remittances from abroad in period t.
$rt(t)$	=	Portion of the carbon tax revenues redistributed to households in period t.[7]
$ts_h(t)$	=	Share of household h in total government transfers in period t.
$ws_h(t)$	=	Share of household h in total wage bill in period t.

---

[7] If $rt(t) = 1$, the carbon tax revenues are fully redistributed as lump-sum transfer on per capita basis to households, whereas, if $rt(t) = 0$, the carbon tax revenues are totally retained in treasury coffers. If $rt(t)$ is somewhere in between, this means that a part of the carbon tax revenues is redistributed to households, with the rest being kept by the government.

Denoting the household income tax rate by $htax_h$, the household disposable income is determined by

$$YD_h(t) = (1-htax_h) \cdot Y_h(t)$$

where

$YD_h(t)$ =   Disposable income of household h in period t.

5.4.5 Government revenues

Total government revenues are generated from four sources: 1) tariffs; 2) indirect taxes; 3) direct taxes and levies; and 4) net foreign borrowing. In the context of CO_2 emissions, an additional source of government revenues is the proceeds of tax levied on carbon emissions. Thus, total government revenues can be expressed as follows:

$$YG(t) = TARIFF(t) + INDT(t) + HHTAX(t) + ETAX(t)$$

$$+ ER(t) \cdot \overline{FBOR}(t) + (1-rt(t)) \cdot RC(t)$$

where

$YG(t)$ =   Total nominal government revenues in period t.
$HHTAX(t)$ =   Total household tax revenues in period t.
$\overline{FBOR}(t)$ =   Net foreign borrowing in US dollars in period t.

$RC(t)$ is discussed in Section 5.8. $TARIFF(t)$, $NETSUB(t)$, $INDT(t)$, $HHTAX(t)$ and $ETAX(t)$ are in turn determined as:

$$TARIFF(t) = \sum_i tm_i \cdot \overline{PWM}_i(t) \cdot ER(t) \cdot M_i(t)$$

$$NETSUB(t) = \sum_i te_i \cdot PWX_i(t) \cdot ER(t) \cdot X_i(t)$$

$$INDT(t) = \sum_i itax_i \cdot PS_i(t) \cdot Q_i(t)$$

$$HHTAX(t) = \sum_h htax_h \cdot Y_h(t)$$

$$ETAX(t) = etx(t)(\sum_i YK_i(t) - DEPR(t) + \overline{GENT}(t))$$

where

$etx(t)$ =   Enterprise tax rate in period t.

5.5   Expenditures

The expenditure block determines the demands for goods and services by households and the government.[8]

5.5.1 Household consumption

In deriving household consumption, I start with the general case where there are different consumption groups and where the consumption categories differ from the sectoral classification of production.

It is assumed that for each household group one single representative consumer allocates a fraction of his income among consumer goods on the basis of the extended linear expenditure system (ELES), whereas the rest is saved and takes the form of purchases of capital goods. The choice for the ELES was essentially motivated by the fact that such system has been estimated on Chinese data (Li, Yang and He, 1985). The ELES accounts for the consumption-saving choice while at the same time allowing for different income elasticities across consumer goods. Following Howe (1975), the ELES is derived from an atemporal maximization of a Stone-Geary function subject to the budget constraint by treating saving as an additional good with zero subsistence quantity. In this case, it takes the form

$$C_{hi}(t)/\overline{POP}_h(t) = \gamma_{hi} + \beta_{hi}/PC_i(t)\left(\frac{YD_h(t)}{\overline{POP}_h(t)} - \sum_j PC_j(t)\gamma_{hj}\right)$$

$$HSAV_h(t)/\overline{POP}_h(t) = (1-\sum_i \beta_{hi})\left(\frac{YD_h(t)}{\overline{POP}_h(t)} - \sum_j PC_j(t)\gamma_{hj}\right)$$

where

$C_{hi}(t)$	=	Total consumption of consumer good i by household h in period t.
$\overline{POP}_h(t)$	=	Total population of type h in period t.
γ_{hi}	=	Per capita subsistence quantity of consumer good i for household h.
β_{hi}	=	Marginal budget share associated with consumer good i for household h.
$HSAV_h(t)$	=	Total nominal savings by household h in period t.

Given the exogenously-specified average budget shares and income elasticities of expenditures on good $i (\zeta_{hi})$, the marginal budget shares can be determined by

---

[8] The determination of demands for new investment in fixed capital and inventories are discussed in the next block.

$$\beta_{hi} = \left(\frac{PC_i(t) \cdot C_{hi}(t)}{YD_h(t)} \right) \zeta_{hi}$$

Total household consumption is the sum of consumer goods. In the ELES specification, it is treated as endogenous. Defining the marginal propensity of household h to consume - $\vartheta_h = \sum_i \beta_{hi}$ - and subtracting savings from disposable income, the following expression for the value of total household consumption can be derived:

$$\sum_h \sum_i PC_i(t) C_{hi}(t) = \sum_h \left(\vartheta_h YD_h(t) + (1 - \vartheta_h) \sum_i PC_i(t) \gamma_{hi} \overline{POP}_h(t) \right)$$

Once the consumption of consumer goods and services is determined, it can be translated into a demand for intermediate goods through a transformation matrix, which defines the contribution of each producing sector to the composition of each of the final consumer goods and services and is also used to compute prices of consumer goods from producer prices (see Section 5.3.8):[9]

$$CI_{hi}(t) = \sum_j tr_{ij} C_{hj}(t)$$

where
$CI_{hi}(t)$ = Demand for intermediate good i by household h in period t.
tr_{ij} = Amount of intermediate good i required to produce one unit of consumer good j, with $\sum_i tr_{ij} = 1$.

5.5.2 Government purchases

As usual, the aggregate real government purchases are set exogenously, and are divided among goods by exogenously-fixed expenditure composition coefficients. Specifically,

$$\overline{G}_i(t) = e_i(t) \cdot \overline{GN}(t)$$

where
$\overline{GN}(t)$ = Aggregate real government purchases in period t.

---

[9] When the consumption categories are the same as the sectoral classification of production, the transformation matrix becomes a unit matrix, with the diagonal elements being equal to unity and the rest to zero.

$e_i(t)$ = Expenditure share of commodity i in total government consumption in period t, with $\Sigma_i \, e_i(t) = 1$.

5.6 Investment and capital accumulation

In this section, I will discuss how to allocate the aggregate investment across sectors, to determine the demand for fixed investment goods by sector of origin and to calculate the total value of depreciation of fixed capital stock.

5.6.1 Allocation of investment across sectors

Section 5.10.2 will discuss how to determine the aggregate nominal gross investment. Once it is determined, the next step is to allocate the aggregate investment across sectors.

In this study, two kinds of capital investments, that is, fixed and circulating capital investments, are distinguished. Circulating capital investment is the so-called inventory investment.

Following Robinson *et al.* (1990), circulating capital investment by sector of destination is determined by exogenous inventory coefficients times production, the former of which defines the amount of circulating capital goods required as inventory in period t in order to produce one unit of good i in that period.

$$SK_i(t) = ac_i(t) \cdot Q_i(t)$$

where
$ac_i(t)$ = Requirement of circulating capital per unit of output i in period t.

Total nominal fixed capital investment is computed by subtracting from the aggregate investment available all investments in circulating capital. Hence,

$$DK(t) = INV(t) - \sum_i P_i(t) \cdot SK_i(t)$$

where
$DK(t)$ = Total investment in fixed assets in period t.
$INV(t)$ = Total nominal investment in period t.

Fixed capital investment is assumed to be allocated among sectors of destination according to the exogenously-given share parameters, which sum to one over all sectors. Accordingly, the sectoral investment is determined by

$$DK_i(t) = ak_i(t) \cdot DK(t)/PK_i(t)$$

where

$DK_i(t)$ =   Investment in fixed assets of sector i in period t.

$ak_i(t)$ =   A share of sector i in total fixed capital investment in period t, with $\sum_i ak_i(t) = 1$. The share parameters are fixed within one period, but can be changed over time to reflect the government priorities of sectoral allocation of investable funds.[10]

5.6.2 Demand for investment goods

Once the sectoral investment allocations are determined, investment in fixed assets by sector of destination can be translated into a demand for fixed investment goods by sector of origin through capital composition matrix:

$$FI_i(t) = \sum_j sf_{ij} \cdot DK_j(t)$$

5.6.3 Total value of depreciation

Total value of depreciation of fixed capital stock is simply the sum of sectoral depreciation of fixed capital stock. That is,

$$DEPR(t) = \sum_i \delta_i \cdot PK_i(t) \cdot K_i(t)$$

5.7 Foreign trade

Before turning to treatment of the foreign sector, I discuss briefly the role of foreign trade in the Chinese economy, because the ways trade is dealt with in an economic model depend to some extent on its role.

In the pre-reform era, China, much like other socialist countries, adopted the doctrine of self-reliance. Foreign trade was thought to be of marginal importance. The ideology has changed since the beginning of the economic reforms in 1978. Foreign trade has come to play an increasingly important role in the Chinese economy. As shown in Table 5.1, in 1978, China's total trade (exports plus imports) amounted to US$ 20.6 billion. By 1990, however, this figure had risen to US$ 115.4 billion. Accordingly, its trade participation ratio (ratio of exports

---

[10] There are alternative ways of determining the time-dependent share parameters. For example, the parameters are adjusted over time according to the relative profit rate of each sector compared with the average profit rate in the economy. More specifically, sectors with higher-than-average profit rates will get a larger share of the investment funds than their share in aggregate capital income (cf. Dervis *et al.*, 1982).

plus imports to GDP) has trebled, from 10.2% to 31.6% over the same period. Its trade with the rest of the world now exerts a strong direct influence on growth and has a profound indirect effect on the modernization of the economy through the impact of technology, marketing skills, as well as contacts with other countries (World Bank, 1990a). Clearly, China is no longer an isolated economy and is becoming an important member of the world trading community.

Table 5.1 *China's exports and imports 1978-1990*

	1978	1983	1984	1985	1986	1987	1988	1989	1990
(billions of US dollars)									
Exports	9.7	22.2	26.1	27.3	30.9	39.4	47.5	52.5	62.1
Imports	10.9	21.4	27.4	42.3	42.9	43.2	55.3	59.1	53.3
Total trade	20.6	43.6	53.5	69.6	73.8	82.6	102.8	111.6	115.4
(as percentage of GDP)									
Exports	4.8	8.0	8.9	10.4	11.0	13.0	12.6	12.4	17.0
Imports	5.4	7.7	9.5	13.5	14.3	13.3	14.6	14.0	14.6
Total trade	10.2	15.7	18.4	23.9	25.3	26.3	27.2	26.4	31.6
(as percentage of world trade)									
Exports	0.80	1.33	1.47	1.52	1.56	1.69	1.77	1.70	1.95
Imports	0.87	1.23	1.41	2.25					1.59

Sources: State Statistical Bureau (1992a); World Bank (1988b, 1990a, 1992a); Singh (1992).

The rising importance of foreign trade suggests a more flexible treatment rather than a very simplistic specification of foreign trade in order to capture the empirical reality of two-way trade.[11] In applied general equilibrium models, the

---

[11] The World Bank (1985b) constructed a dynamic input-output model to analyse long-term patterns of growth and structural change in the Chinese economy over the period 1981-2000. In the modelling exercise, foreign trade was specified in a very simplistic manner so that foreign trade in a given sector adjusted merely to fill the gap between domestic demand and supply in each period. That is, if demand for domestically-produced goods is greater than supply, imports will fill this gap and exports will be fixed at a pre-determined ratio of gross output. Conversely, if demand for domestically-produced goods is smaller than supply, exports will fill the gap and imports will be fixed at a pre-determined ratio of domestic demand (cf. World Bank, 1985a). The simplistic treatment of China's trade is also shown in the study of Ezaki and Ito (1993), who examined the impact of market liberalization in China through a CGE approach. In their study, the domestically-produced goods are assumed to be perfect substitutes for the imported goods when modelling foreign transactions.

flexibility is usually attained by incorporating the so-called Armington specification (Shoven and Whalley, 1984), which means that the imports originating from the rest of the world are imperfect substitutes for the domestically-produced goods. Therefore, the present CGE model of the Chinese economy also adopts the concept of product differentiation.[12] Accordingly, the model has relative-price-dependent import demand and export supply functions, reflecting the choices by agents at home and abroad between the imported and domestically-produced goods. This treatment of imports and exports partially insulates the domestic price system from changes in world prices (Robinson *et al.*, 1990). In addition, I also impose the small-country assumption, which implies that the world (dollar) prices of imports and exports are considered to be exogenously determined. The detailed specification is as follows.

5.7.1 Import demand

In the classical theory of trade it is assumed that domestically-produced goods are perfect substitutes for imported goods. For a country such as China, this assumption is troublesome. First, quality differences are frequently observed between imports and domestic substitutes. Second, at a level of aggregation of ten sectors, the goods are fairly aggregated, so product differentiation does exist in the same sector. In our model, we solve this problem by incorporating the Armington specification. Thus, domestic users are taken to use a composite commodity i, which is a CES aggregate of imported and domestically-produced goods of type i:

$$Z_i(t) = \Psi_i \left(\mu_i M_i(t)^{-\varphi_i} + (1-\mu_i) D_i(t)^{-\varphi_i} \right)^{-1/\varphi_i}$$

where

$Z_i(t)$ =   Total supply of composite commodity i in period t.

φ_i = $(1-\psi_i)/\psi_i$, where ψ_i is the price elasticity of substitution between imported and domestically-produced commodity i.

Given the imperfect substitutes between the goods from two sources of supply, the domestic good price $PD_i(t)$ and the domestic price of imports $PM_i(t)$, the

---

[12] The implication of the choice is that a substantial investment is required in data gathering, because there is only one single column for net export in the current Chinese input-output table and hence imports and exports in gross terms need to be estimated for this study. See Section 6.3 for a further discussion.

domestic users are assumed to minimize total expenditure on composite commodity i:[13]

$$\min_{M_i(t),D_i(t)} \quad [PM_i(t)M_i(t) + PD_i(t)D_i(t)]$$

subject to the CES transformation function defined above. Thus, the first-order conditions for expenditure minimization yield the following optimal import demand:

$$\frac{M_i(t)}{D_i(t)} = \left(\frac{\mu_i}{1-\mu_i}\right)^{\psi_i} \left(\frac{PD_i(t)}{PM_i(t)}\right)^{\psi_i}$$

This import demand specification implies that the domestic users will choose a mix of $M_i(t)$ and $D_i(t)$, depending on their relative prices. The degree to which the domestically-produced goods may be substituted for the imported goods is measured by the price elasticity of trade substitution ψ_i. In the classical theory of trade, ψ_i is infinity, so that $PD_i(t)=PM_i(t)$, since if $PD_i(t)$ ever exceeded $PM_i(t)$, $D_i(t)$ would have to be zero. This is the least realistic case where the domestically-produced goods are perfect substitutes for the imported goods. In the case where $\psi_i=0$, the demand ratio is fixed, so the goods from two sources of supply are perfect complements. In the Cobb-Douglas case where $\psi_i=1$, the ratio of cost shares is independent of price changes.[14] The three cases can be considered to be special cases of the demand specification above. In all other cases but the special cases, as ψ_i gets larger, the sensitivity of the demand ratio $(M_i(t)/D_i(t))$ to changes in the relative prices rises.

5.7.2 Export demand

Being a small country on the export side means that domestic exports only constitute a small fraction of the world market for that product, so it may not be able to affect the world market price with its exports. Incorporating the product differentiation effect, however, a country can certainly increase its market share in a certain product by lowering the dollar price of that product. To illustrate

---

[13] Total expenditure on composite commodity i by domestic users is determined by

$$P_i(t)Z_i(t) = PM_i(t)M_i(t) + PD_i(t)D_i(t)$$

[14] This is clearly shown when multiplying the demand ratio $(M_i(t)/D_i(t))$ by the corresponding price ratio $(PM_i(t)/PD_i(t))$.

this, I assume that the world demand functions for Chinese products take the following simple constant elasticity form:

$$X_i(t) = xo_i e^{grx_i(t)*(t-t_o)} \left(\frac{\overline{PWO}_i(t)}{PWX_i(t)} \right)^{\xi_i}$$

where

xo_i	=	Base year exports of commodity i.
$\overline{PWO}_i(t)$	=	World (dollar) price of exports of commodity i in period t.
$grx_i(t)$	=	*ex ante* export growth rate of commodity i in period $(t-t_o)$.
ξ_i	=	Price elasticity of export demand for commodity i.

Notice that the export demand for Chinese products specified in this manner may exhibit a strong response to changes in the dollar prices of their own products (cf. Dervis *et al.*, 1982). This may be not realistic. Moreover, the specification requires information on the artificial world composite prices and the substitution elasticities between the differentiated products on the world market (cf. Peerlings, 1993). For these reasons, the specification of export demand has been eliminated from this model.[15] Instead, the export supply functions described below are utilized to determine the exports of Chinese products.

5.7.3 Export supply

On the export side, the assumption of product differentiation is reflected in total domestic production, which is assumed to be a constant elasticity of transformation (CET) aggregation function of exported goods and domestically-consumed goods. The aggregation is given by

$$Q_i(t) = \Phi_i \left\{ v_i X_i(t)^{\phi_i} + (1-v_i) D_i(t)^{\phi_i} \right\}^{1/\phi_i}$$

where

Φ_i	=	Shift parameter associated with gross output of sector i in export supply function.
v_i	=	Share parameter associated with exported good i in export supply function.
ϕ_i	=	$(1+\eta_i)/\eta_i$, where η_i is the price elasticity of transformation between foreign and domestic sales of commodity i.

---

[15] De Melo and Robinson (1989) and Garbaccio (1994) also dropped the export demand specification in their CGE models.

Given the domestic good price $PD_i(t)$ and the domestic price of exports $PX_i(t)$, the domestic producers are assumed to maximize total profit from sales of good i:[16]

$$\min_{X_i(t),D_i(t)} \quad [PX_i(t)X_i(t) + PD_i(t)D_i(t)]$$

subject to the CES transformation function defined above. Thus, the first-order conditions for profit maximization yield

$$\frac{X_i(t)}{D_i(t)} = \left(\frac{1-v_i}{v_i}\right)^{\eta_i} \left(\frac{PX_i(t)}{PD_i(t)}\right)^{\eta_i}$$

This export supply function clearly indicates that exports depend on the relative prices of exported goods. The degree to which the domestically-produced goods may be substituted for the exported goods is measured by the elasticity of trade transformation. Similar to the discussion in Section 5.7.1, if η_i is infinity, the domestically-produced goods are perfect substitutes for the exported goods; if $\eta_i=0$, the supply ratio is fixed, so the goods of two types are perfect complements; and if $\eta_i=-1$, the ratio of revenue shares is independent of price changes.[17] The three cases are special cases of the supply specification above. In all other cases but the special cases, if η_i is positive and large, a small change in the relative prices has a large impact on the supply ratio $(X_i(t)/D_i(t))$.

5.7.4 Foreign trade deficit

Foreign trade deficit in US dollars is defined as

$$FTD(t) = \sum_i \overline{PWM}_i(t) \cdot M_i(t) - \sum_i PWX_i(t) \cdot X_i(t)$$

where
$FTD(t)$ =   Foreign trade deficit in US dollars in period t.

---

[16] Total profit from sales of good i by domestic producers is determined by

$$PS_i(t)Q_i(t) = PX_i(t)X_i(t) + PD_i(t)D_i(t)$$

[17] This is clearly shown when multiplying the supply ratio $(X_i(t)/D_i(t))$ by the corresponding price ratio $(PX_i(t)/PD_i(t))$.

5.8 Energy and Environment

In this section, I first discuss how to incorporate the possibilities of decoupling energy consumption and CO_2 emissions from GDP growth into the CGE model. Next, I present the way in which the energy consumption and the corresponding CO_2 emissions are calculated.[18] Then I introduce the carbon tax as an economic instrument designed for limiting CO_2 emissions. Finally, I provide the way in which to analyse the contribution of changes in both level and structure of economic activity, a change in energy input coefficients, and a change in direct energy consumption by households to energy consumption reduction as a result of the imposition of carbon taxes.

5.8.1 Decoupling energy consumption and CO_2 emissions from GDP growth

Energy consumption and CO_2 emissions are closely linked to the GDP, but need not grow at the same rate as GDP. In the CGE model, the possibilities of decoupling energy consumption and CO_2 from GDP growth are represented by two macroeconomic parameters. One is termed AEEI (autonomous energy efficiency improvement). The AEEI parameter accounts for all but energy price-induced energy conservation. Energy conservation of this type is available at zero or negative net cost. In cost-benefit analysis of greenhouse gas control, this implies, *ceteris paribus*, a higher optimal level of emission reduction than when abatement costs are always positive (cf. Ayres and Walter, 1991). Energy conservation of this type is taking place regardless of the development of energy prices. It may be brought about by deliberate changes in public policy, for example, efficiency standards and various utility programmes. It may also occur as a result of 'good housekeeping' or of a shift in the economic structure away from energy intensive heavy manufacturing towards less energy intensive services (Williams, 1990; Manne and Richels, 1992). In the case where AEEI lowers the rate of growth of CO_2 emissions over time, and therefore decreases the amount by which CO_2 emissions need to be constrained, the economic impacts of a given carbon constraint will also be lower. The second parameter is known as the elasticity of price-induced substitution between the inputs of capital, labour and energy. It serves as a factor measuring the ease or difficulty of substitution for energy during a period of rising energy prices. Thus, energy conservation of the second type occurs as a reaction to rising energy prices.

---

[18] Only CO_2 emissions from the burning of fossil fuels are calculated; those from deforestation are not taken into account.

5.8.2 Computing energy consumption and CO_2 emissions[19]

For simplicity, our CGE model is normalized in such a way that all prices are equal to one in the base year. As a result, the underlying volumes or quantities are equal to the corresponding nominal output values in the base year and do not correspond directly to physical units. Thus in order to calculate the amount of CO_2 emissions, real energy consumption per sector is translated into physical terms, using energy-specific technical conversion factor, and then the physical terms are converted into CO_2 emissions, using energy-specific emission coefficients. A similar procedure is applied to the calculation of CO_2 emissions by households. These coefficients are then used to compute CO_2 emissions in each simulation.

5.8.3 Carbon tax as a means of limiting CO_2 emissions

Domestic CO_2 emissions can be reduced by means of emission standards (that is, command-and-control regulations), energy taxes or carbon taxes. Emission standards are imposed on one or more branches of industry, which are often the most polluting in terms of CO_2 emissions. According to the theory of environmental economics, emission standards lead to higher costs than economic instruments in order to achieve the same reduction of CO_2 emissions. This has been confirmed in the study of Beauséjour *et al.* (1995), the results of which show that Canada's GDP loss in 2000 from imposing emission standards on four industries (iron and steel, electric utilities, transportation and services) is 33% greater than from an energy tax in order to stabilize Canada's CO_2 emissions at 1990 levels in the year 2000. This implies that an energy tax is more cost-effective in meeting an emission target than emission standards. In Section 2.5, however, it has been argued that a carbon tax is even more cost-effective in terms of target achievement than an energy tax. Moreover, compared with an energy tax, a carbon tax is less burdensome in that it raises a smaller amount of government revenues for a given reduction of CO_2 emissions.[20] For these reasons, our CGE model incorporates a carbon tax as a means of achieving the pre-defined target of CO_2 emissions.

---

[19] This procedure of computing energy consumption and CO_2 emissions has also been used in the OECD GREEN model (Burniaux *et al.*, 1992) and the DGEM model (Jorgenson and Wilcoxen, 1993a, 1993b).

[20] Any government revenue is in itself a source of inefficiency, since there is no practical way to raise revenue through taxes without causing some distortion. Clearly, the finding is in accordance with this basic economic theory and is shown in the empirical studies of Jorgenson and Wilcoxen (1993b) and Beauséjour *et al.* (1995).

A carbon tax is an excise tax, which is expressed as a fixed amount of Chinese currency per ton of CO_2 emissions. It applies to the consumption of primary fuels only.[21] Consequently, the energy sector only pays carbon taxes on its own use of fuels. Rather than computing optimal carbon taxes, the CGE model can be used to simulate the effects of arbitrary carbon taxes, or to calculate the taxes that are required to achieve the pre-defined emission target. In the latter case where an upper bound on total CO_2 emissions is imposed, the carbon tax is viewed as the equilibrium shadow price associated with the emission constraint.

Given the carbon tax, it is possible to generate the following government revenue from the proceeds of this tax:[22]

$$RC(t) = \sum_{i=1}^{10} \sum_{j=7}^{10} tc \cdot \varepsilon_j \cdot \theta_j \cdot VE_{ji}(t)$$

where

tc = Carbon tax, expressed as a fixed amount of Chinese currency per ton of carbon emissions.

ε_j = CO_2 emission coefficients of fuel j.

θ_j = Factor converting real consumption of fuel j into physical terms, for example, Terajoules.

Given the government revenues by kind of fuel j, a given carbon tax can be converted into fuel-specific *ad valorem* tax rate, $\tau f_j(t)$, which is defined to be the ratio of government fuel-specific revenues to the total values of domestic absorption of the fuel as follows

$$\tau f_j(t) = \frac{tc \cdot \varepsilon_j \sum_{i=1}^{10} \theta_j \cdot VE_{ji}(t)}{PD_j(t) \cdot D_j(t) + PM_j(t) \cdot M_j(t) - PD_j(t) \cdot X_j(t)}$$

5.8.4 Contribution to energy consumption reduction

The imposition of a carbon tax will reduce total energy consumption. In order to highlight the operation of the different adjustment mechanisms, total energy

---

[21] The distinction between a production and a consumption based carbon tax would affect the assessments of international incidence. These issues are discussed in Section 2.5, Whalley (1991), and Whalley and Wigle (1991a, 1991b).

[22] This means that all sectors are taxed. Alternatively, only certain sectors are taxed. According to the study of Conrad and Schröder (1991), this option is more costly than the former in terms of GDP forgone because restricting the tax base reduces substitutions in energy use relative to the 'tax everyone' case.

consumption reduction with the carbon constraint relative to the baseline scenario
is subdivided into four components following Bergman (1988) and Bergman and
Lundgren (1990). Thus we have the following identity

$$TOT(t) = VOL(t) + COMP(t) + INP(t) + DIR(t)$$

where

$TOT(t)$	=	Total change in energy consumption with and without the carbon constraint in period t.
$VOL(t)$	=	Change in energy consumption due to a change in aggregate production in period t, provided that aggregate production is composed in the same way as under the baseline scenario.
$COMP(t)$	=	Change in energy consumption due to a change in composition of aggregate production in period t.
$INP(t)$	=	Change in energy consumption due to a change in energy input coefficients in period t.
$DIR(t)$	=	Change in energy consumption due to a change in direct energy consumption by households in period t.

In this identity, $TOT(t)$, $VOL(t)$, $COMP(t)$, $INP(t)$ and $DIR(t)$ are, in turn, determined as follows:

$$TOT(t) = EDEM^c(t) - EDEM^b(t)$$

$$VOL(t) = \sum_{i=1}^{10} eo^b_i(t)[QQ_i(t) - Q^b_i(t)]$$

$$COMP(t) = \sum_{i=1}^{10} eo^b_i(t)[Q^c_i(t) - QQ_i(t)]$$

$$INP(t) = \sum_{i=1}^{10} [eo^c_i(t) - eo^b_i(t)]Q^c_i(t)$$

$$DIR(t) = \sum_{i=1}^{10} [ED^c(t) - ED^b(t)]$$

where

$EDEM^b(t)$	=	Total energy consumption under the baseline scenario in period t.
$EDEM^c(t)$	=	Total energy consumption with the carbon constraint in period t.
$ED^b(t)$	=	Direct energy consumption by households under the baseline scenario in period t.

$ED^c(t)$ =   Direct energy consumption by households with the carbon constraint in period t.

$Q^b_i(t)$ =   Gross output of sector i in period t under the baseline scenario.

$Q^c_i(t)$ =   Gross output of sector i in period t with the carbon constraint.

$QQ_i(t)$ =   Gross output of sector i in period t with the carbon constraint, provided that aggregate production is composed in the same way as under the baseline scenario. It is calculated as $[Q^b_i(t)/\Sigma_i Q^b_i(t)] \times \Sigma_i Q^c_i(t)$.

$eo^b_i(t)$ =   Energy-output ratio of sector i in period t under the baseline scenario.

$eo^c_i(t)$ =   Energy-output ratio of sector i in period t with the carbon constraint.

In the identity above, energy types are not specified. This implies that the decomposition identity can be applied to any energy type: coal, oil, natural gas, or electricity. Needless to say, this identity provides a very useful way to analyse the contribution of each adjustment mechanism to energy consumption reduction.

5.9 Welfare measures

As in many applied general equilibrium models, our model also takes the Hicksian equivalent variation as a measure of the welfare impacts of emission abatement policies.[23] Equivalent variation takes the pre-policy equilibrium income and consumer prices as given and measures the changes in income required to obtain post-policy utility level at pre-policy consumer prices.[24] This can be written as follows:

$$EV(t) = E(U^s(t), PC^b(t)) - E(U^b(t), PC^b(t))$$
$$= \sum_i PC^b_i(t) \sum_h C^s_{hi}(t) - \sum_i PC^b_i(t) \sum_h C^b_{hi}(t)$$

---

[23] Equivalent variation is a convenient choice because it measures the income change at pre-policy prices. This makes equivalent variation more suitable for comparisons among a variety of policy changes compared with the compensating variation. See Boadway and Bruce (1984) and Varian (1992) for a further discussion.

[24] For this definition of equivalent variation, see Shoven and Whalley (1992). For an alternative definition, see Varian (1992).

where

$EV(t)$	=	Equivalent variation in period t.
$E(U^s(t), PC^b(t))$	=	Expenditure function[25] that gives the expenditure necessary to achieve post-policy utility level $U^s(t)$ at pre-policy consumer prices $PC^b(t)$ in period t.
$E(U^b(t), PC^b(t))$	=	Expenditure function that gives the expenditure necessary to achieve pre-policy utility level $U^b(t)$ at pre-policy consumer prices $PC^b(t)$ in period t.
$PC^b(t)$	=	Vector of pre-policy consumer prices in period t.
$PC^b_i(t)$	=	Pre-policy consumer price of good i in period t.
$C^b_{hi}(t)$	=	Pre-policy consumption of consumer good i by household h in period t.
$C^s_{hi}(t)$	=	Post-policy consumption of consumer good i by household h in period t.

At each point in time, $EV(t)$ can be calculated according to the equation above. If $EV(t)$ is positive, post-policy welfare is improving; if negative, post-policy welfare is worsening.[26]

5.10   Market clearing conditions and macroeconomic balances

The last block defines the market clearing conditions and macroeconomic balances that the model must satisfy.

5.10.1 Market clearing conditions

Our CGE model is of a time-recursive dynamic structure. The agents modelled are assumed to behave myopically, thus reacting to current prices only. Consequently, the economy evolves in a sequence of period-related, but intertemporally uncoordinated, temporary flow equilibria as compared with an intertemporal equilibrium under the assumption of perfect foresight (Pereira and Shoven, 1988; Stephan, 1992; Gunning and Keyzer, 1995). This temporary

---

[25] For the definition of an expenditure function, see Varian (1992).

[26] Since expenditure does not include the value of leisure, our welfare measures do not reflect all the conventional welfare aspects. A completely conventional measure should be based on full consumption, which consists of expenditure on non-durables, services of durables, and the value of leisure. In the context of 'greening' the conventional measure, account should also be taken of the consumption value of environmental services. See, for example, Dasgupta *et al.* (1995) for a detailed discussion of conventional and 'green' national accounts.

equilibria are taking place in each period such that the market clearing conditions for commodities and primary factors are satisfied.

5.10.1.1 Product market clearing

Product market clearing implies that the sectoral supply of composite commodities must equal all domestic demands. Specifically,

$$Z_i(t) = V_i(t) + \sum_k CI_{hi}(t) + \overline{G}_i(t) + FI_i(t) + SK_i(t)$$

This condition specifies that the total supply of each composite commodity must equal the sum of intermediate demands, consumption demands by households and government, and investment demands for the composite commodity in the same category.

5.10.1.2 Labour market clearing

The standard assumption that labour is viewed as homogeneous and mobile across sectors in response to changes in the demands for labour is used in this model (Dervis *et al.*, 1982; Martin, 1990). Full employment[27] and supply of labour being wage-inelastic are also assumed (Peerlings, 1993). Thus, in each period, the labour market clears when total labour demand by sectors, which is simply the sum of endogenously-determined demands for labour in each sector, is equal to the exogenously-projected supply of total labour force

$$\sum_i L_i(t) = \overline{LS}(t)$$

where

$\overline{LS}(t)$ = Total labour force available in period t.

5.10.1.3 Capital market clearing

In the short run, capital is sector-specific. Therefore, unlike the single clearing condition for labour market given above, for capital there is a separate market clearing equation for each sector in each period. This reflects the assumption that, even in the short run, labour is mobile across sectors whereas capital, once

---

[27] An alternative to the full employment with flexible wage rate would be to fix the wage rate exogenously and to let the model determine the employment level (Dixon *et al.*, 1982). But this specification would not be in line with the *general equilibrium* assumption. Moreover, according to the study of Proost and Regemorter (1994), in the context of reducing CO_2 emissions, the specification is much more costly than the former in terms of GDP forgone because fixing the wage rate reduces substitutions between labour and energy relative to the flexible wage rate case.

installed in a sector, is immobile. However, in the long run, capital is assumed to be intersectorally mobile. Thus, given the exogenously-determined amount of total fixed capital in each period, the capital market clearing condition, which is similar to that of labour market, simply becomes

$$\sum_i K_i(t) = \overline{FK}(t)$$

where

$\overline{FK}(t)$ =   Total capital stock available in period t.

5.10.2 Macroeconomic balances

In our CGE model, the macroeconomic balances can be seen as describing macroeconomic equilibrium conditions for the government budget, balance of trade, and savings-investment balance. Moreover, since the model is closed in that it satisfies Walras' Law (cf. Varian, 1992), the three macroeconomic balances must satisfy the identity: private savings + government savings + foreign savings $\equiv$ aggregate investment.

I begin with the government budget. In each period, government revenues are assumed to balance with total government expenditures that include government purchases, savings, transfer payments to enterprise and households, and export subsidies. Thus the government revenue-expenditure balance takes the form

$$YG(t) = \sum_i [P_i(t) \cdot \overline{G}_i(t)] + GSAV(t) + \overline{GENT}(t) + \overline{HHT}(t) + NETSUB(t)$$

where

$GSAV(t)$ =   Total nominal government savings in period t. It serves as the equilibrating variable to keep the balance.

As far as the surplus on balance of trade is concerned, it is determined as the sum of net foreign borrowing, net remittances from abroad, net foreign savings and less foreign trade deficit:[28]

$$\overline{SBT}(t) = \overline{FBOR}(t) + \overline{REMIT}(t) + \overline{FSAV}(t) - FTD(t)$$

where

$\overline{SBT}(t)$ =   A surplus on balance of trade in period t.

$\overline{FSAV}(t)$ =   Net foreign savings in US dollars in period t.

---

[28] Net foreign borrowing refers to net foreign transfers. Unlike net foreign savings that are entered as payment to the capital account from the rest of world, net foreign borrowing is distributed among various actors.

The model takes the surplus on balance of trade to be exogenous.[29] Also net foreign borrowing, remittances from abroad and net foreign savings are set exogenously. The effect is that the exchange rate serves as the equilibrating variable to ensure the balance of trade constraint. Alternatively, the exchange rate can be fixed (relative to the price numéraire), with the balance of trade and either net foreign borrowing or net foreign savings becoming endogenous.

Now I consider the savings-investment balance. This relates to what is known as the macro closure of CGE models. The model operates with the 'neoclassical closure'.[30] The choice for the closure rule is motivated by the World Bank's forecast that high rates of savings, which have characterized the Chinese economy, are for a number of reasons likely to continue in the medium-term future (cf. World Bank, 1990d). The rule requires that total savings equals aggregate gross investment. The former is defined as the sum of savings by households, government and enterprises, depreciation and net foreign savings:

$$TSAV(t) = \sum_h HSAV_h(t) + GSAV(t) + ESAV(t) + DEPR(t) + ER(t) \cdot \overline{FSAV}(t)$$

where
$TSAV(t)$ = Total nominal savings in period t.

With savings by households and enterprises and depreciation determined by fixed savings and depreciation rates, government savings partially determined by fixed tax rates, and foreign savings set exogenously, the net effect is that our model is a savings-driven model.

As discussed above, the three macroeconomic balances are not all independent, in accordance with Walras' Law. One of them should be dropped in order to find a solution. In our model, we drop the savings-investment balance.[31]

---

[29] Similar to the USDA/ERS CGE model (Robinson *et al.*, 1990), our model is based on the GNP accounts, so trade in services includes factor services. Consequently, the balance of trade in our model is the balance on current account.

[30] There is extensive literature on alternative macro closures of CGE models. The diversity of closure rules reflects the different theoretical views of how the macroeconomic system works. See for example Dewatripont and Michel (1987), Robinson (1989), and Robinson *et al.* (1990) for a further discussion.

[31] When the system is over-determined and one of the constraints of the CGE model must be relaxed in order to find a solution, choosing a particular closure rule means precisely deciding which constraint should be dropped (cf. Dewatripont and Michel, 1987).

This completes the description of our CGE model. In the next section, the ways the model can be used will be described and some concluding remarks with respect to directions for further work will be drawn.

5.11   Concluding remarks: directions for further work

The above-described CGE model makes it possible to analyse the Chinese economy-energy-environment system interactions simultaneously. While detailed application of this model will follow in Chapter 7, this section summarizes the ways in which the model can be used.

First, to make conditional forecasts for economic development, energy consumption and CO_2 emissions in China under a number of the exogenous assumptions. In the absence of specific policy aimed at CO_2 control, such a forecast is usually labelled as the baseline scenario or business-as-usual scenario.

Second, to carry out the counterfactual policy simulations. The counterfactual scenarios can be specified in terms of the pre-defined carbon limits or carbon taxes with and/or without the carbon tax revenue recycling options. Such counterfactual simulations allow us to compute the implications of alternative carbon limits relative to the business-as-usual scenario, and the efficiency improvement of carbon tax revenue recycling scenarios relative to the carbon tax retention ones.

Third, to undertake sensitivity analysis. This is mainly to assess the robustness of the simulation results to the key parameter values chosen when developing the baseline scenario and counterfactual scenarios.

Clearly, the CGE model provides a suitable and flexible basis for analysing the economic impacts of compliance with CO_2 emission limits. Nevertheless, there are some areas where there is a need for further methodological and empirical work to enrich the policy relevance of the current CGE model.

5.11.1 Incorporating an intertemporal optimization structure

The current CGE model is of a time-recursive dynamic structure. The agents modelled are assumed to behave myopically, thus reacting to current prices only. Consequently, the economy evolves in a sequence of period-related, but intertemporally uncoordinated, temporary flow equilibria. This suggests that further modelling work is required on incorporating intertemporal optimization decision making into the current model by adopting the assumption of perfect foresight, so energy producers and consumers will be able to sufficiently foresee the scarcities of energy and the environmental restrictions that would be developed during the coming decades.[32]

This change in behavioural assumptions and hence in the model structure is of policy relevance. In the CO_2 context, the assumption of myopic behaviour implies that CO_2 emission constraints are imposed in the form of annual ceilings, so the emission profile over time is completely fixed. For at least two reasons, this myopic assumption is not very realistic in terms of credible government commitments to curb CO_2 emissions. First, greenhouse gases are the so-called 'stock pollutants'. The greenhouse effect is not caused by the flow of emissions as such, but by their accumulation in the atmosphere. Thus, the effectiveness of abatement policies should be judged in terms of a reduction in accumulated CO_2 emissions during the period under consideration, rather than in terms of the emissions rate at any particular point in time. Second, imposing a cumulative carbon limit provides an additional degree of policy flexibility, since it allows a country to optimize its time path of carbon emissions. If total carbon emissions with a cumulative carbon limit are set the same as those with an annual carbon limit, emission reductions occur later with a cumulative limit than they would with an annual limit. This is because the economic costs of carbon abatement are minimized if a reduction in CO_2 emissions occurs later. There are two major reasons for this. First, a cumulative carbon limit allows more time for adjustments of technologies and capital. Second, any consumption losses have a heavier weight in the maximand than losses in the more distant future, simply because the welfare function incorporates discounting considerations. That postponing a reduction in CO_2 emissions would lead to a welfare gain[33] has been confirmed in the study of Blitzer *et al.* (1992), the results of which suggest that economic performance measured by Egypt's GDP under an accumulative emissions constraint is superior to that under an annual emissions constraint.

This change in the model structure is certainly required for linkage between the CGE model and the power planning model presented in Chapter 8. The former is of a recursive structure, the latter of an intertemporal one. Clearly, the structures

---

[32] To develop a CGE model of the Chinese economy in the course of one PhD research, including estimating empirical parameters and carrying out policy simulations, is a formidable challenge (Bruggink, 1995). Thus, given time constraints, I adopt a time-recursive structure because incorporating an intertemporal optimization structure would not only complicate the model but also have the additional empirical requirements.

[33] Within the cost-benefit framework, this conclusion does not always hold, since earlier emissions under a cumulative carbon constraint lead to earlier warming. Whether or not a cumulative limit policy performs better than an annual one will depend on how certain the costs and benefits of emission reductions are. For a further discussion of this issue, see, for example, Peck and Teisberg (1993).

of the two models are not compatible with each other. Thus, the current CGE model needs to be tailored to the linkage with the power planning model in order to obtain meaningful feedback effects of compliance with CO_2 limits in China's electricity sector on the macroeconomic indices.

5.11.2 Improving the modelling of production structure

In the current model, energy inputs, including coal, oil, natural gas and electricity, combine by means of a Cobb-Douglas function into an energy aggregate. This specification imposes a unitary elasticity of substitution between a pair of inputs. Given the difficulties in saving electricity and the increasing advantages of electricity use in new industrial processes, however, there is some evidence that fossil fuels can be substituted with electricity more difficultly than with each other. This consideration suggests that additional nesting levels in the production structure need to be introduced.

Moreover, relationship between capital and energy in the nesting hierarchy is controversial, although energy and labour appear uniformly to be substitutes (Tietenberg, 1992). Empirical estimates suggesting complementarity between the two factors are at least as frequent as findings suggesting substitutability (Burniaux et al., 1992; Tietenberg, 1992). This uncertainty, combined with that energy and labour are often found to be substitutable to the same extent as capital and labour,[34] would support a nesting hierarchy based on labour and a capital/energy bundle, whereas our model, like most models addressing CO_2 issues, has a nesting hierarchy based on energy and a capital/labour bundle. Given that fossil fuel combustion is an important source of CO_2 emissions, whether capital and energy are complements or substitutes is especially important in light of the links between energy use and climate change.[35] Thus, investigating the capital/energy relationship further is worthwhile.

In addition, attention is also paid to incorporating the so-called low-carbon or carbon-free backstop technologies, which is considered to be critical in reducing the costs induced for compliance with the emission limits (see Section 2.5 for a further discussion).

---

[34] This was found in a good survey of the elasticities of substitution between capital and labour, energy and labour, and energy and capital (Burniaux et al., 1992).

[35] Reducing fossil fuel consumption to a large extent is the goal of any strategy for responding to climate change, indicating that if capital is complementary to energy, response strategies would have the side effect of reducing the rate of capital formulation.

5.11.3 Incorporating estimates of the benefits from reduced CO_2 emissions

The current model is confined to the economic impacts of carbon emission limits. This measurement of costs is only part of the story. In order to arrive at an overall judgement, policymakers also need the information on the ecological damages avoided by the policies aimed at limiting CO_2 emissions. Without the estimates on the benefit side of reduced emissions, there could be a bias towards inaction.[36] Clearly, it is necessary to quantify the benefits that CO_2 control brings about, so policymakers can make their decision balancing both the benefits and costs of policy implementation. Unfortunately, information of this type is not sufficiently available at present. If it becomes available at some time in the future, this should be included in the CGE model.

---

[36] See Footnote 2 in Chapter 4.

6 Data, Model Calibration and Solution[1]

6.1 Introduction

Credible economic policy analysis stands on two legs. One is methodology, and the previous chapter has been devoted to advocating computable general equilibrium (CGE) modelling for this purpose. The second fundamental requirement is the availability of reliable data. This component is essential for empirical application of any CGE model, although it is often problematic.

The essential data required for a well-calibrated CGE model can be aggregated into three broad categories: a) detailed economic accounts, which are ideally maintained in the form of a social accounting matrix (SAM); b) structural parameters; and c) a number of subsidiary data (cf. Roland-Holst *et al.*, 1993). The construction of the data base associated with any CGE model requires great efforts, since models of the CGE type are data-hungry. This work is crucial to the quality of models of this type and their results as well as to the scope of their applicability.

In this chapter, I will discuss the data requirements of the CGE model of the Chinese economy and some problems arising from constructing the corresponding data base. Section 6.2 presents the sectoral classification for this study. In Section 6.3, the construction of the social accounting matrix is described. The determination of structural parameters is given in Section 6.4, while subsidiary data are the subject of Section 6.5. In Section 6.6, the solution approaches to finding numerical solutions to CGE models are discussed briefly. Section 6.7 presents some concluding remarks.

6.2 Sectoral classification

Before turning to the construction of the SAM, the sectoral classification for this study is first presented in Table 6.1, together with the corresponding sector codes of the original 33-sector input-output (I-O) table (State Statistical Bureau, 1991) from which the 10-sector is aggregated. The sectoral abbreviation in parenthesises also appears in Table 6.1, which is used to denote the sectors in the coming sections when discussing the sector-related parameters and data.

This sectoral aggregation is mainly dictated by the need to facilitate the analysis of the policy issues that the CGE model addresses, while trying to reduce the

---

[1] This chapter forms the input to the paper in *Economic Systems Research* (Zhang, 1997c).

time spent on both preparation of the data set necessary for running the model
and computation is taken into account. That is to investigate, among other things,
endogenous substitution among energy inputs and alternative allocation of
resources as well as endogenous determination of foreign trade in the national
economy in order to cope with the environmental restrictions, at both sectoral
level and macroeconomic level.

Table 6.1   *Classification of producing sectors in the CGE model*

10-Sector	33-Sector
1. Agriculture (AGRI)	01
2. Heavy industry (HIND)	04,05,14-24
3. Light industry (LIND)	06-10
4. Transport & Communication (TRCO)	26
5. Construction (CONS)	25
6. Services (SERV)	27-33
7. Coal (CIND)	02,13
8. Oil (OIND)	03[a],12
9. Natural gas (NGAS)	03[a]
10. Electricity (ELEC)	11

[a] In the 33-sector input-output table, the natural gas sector is part of the oil sector. It is separated
from the oil sector using the 117-sector input-output table (State Statistical Bureau, 1991).

6.3   Construction of the social accounting matrix

An SAM is the cornerstone of a CGE model. It usually starts with an input-
output table. The first Chinese I-O tables released are for 1981 (State Statistical
Bureau, 1986). These tables were compiled according to the Material Product
System (MPS) and the nonmaterial service sector in these tables excluded many
service activities being concealed within other sectors, especially in the industrial
sector. Also public consumption excluded military investment because the MPS
treats it as a fixed capital investment.[2] The World Bank (1985b) compiled its
own set of tables for 1981 according to both the MPS and the United Nations'
System of National Accounts (SNA). In 1991, the State Statistical Bureau
published a set of tables for 1987. In contrast to previous tables, the 1987 I-O
tables follow the SNA conventions closely, though not exactly. These I-O tables
are aggregated at 6, 33, and 117 sector levels. At the 33-sector level the table

---

[2] The opposite of fixed capital investment is circulating (working) capital
investment that is often called inventory investment.

includes a make (activity by commodity) matrix and a use (commodity by activity) matrix.

In an SAM, each row and column represent the income and expenditure accounts of the various actors, respectively. The respective row and column sums must balance for each account. In equilibrium, this balance implies: (1) costs (plus distributed earnings) exhaust revenues for producers; (2) expenditure (plus taxes and savings) equals income for each actor; and (3) demand for each commodity equals its supply (Robinson *et al.*, 1990). More specifically, for an aggregated 1987 SAM that I have estimated for China following the above-mentioned sector classification, the equilibrium condition implies the following:

Domestic sales for each commodity $\equiv$ Total absorption of that commodity;
Total sales by activity $\equiv$ Total costs of that activity;
Total net value added (row sum) $\equiv$ total net value added (column sum);
Total labour compensation $\equiv$ total household expenditure;
Enterprise income $\equiv$ enterprise expenditure;
Government income $\equiv$ government expenditure;
Total saving $\equiv$ total investment;
Foreign trade losses (row sum) $\equiv$ foreign trade losses (column sum);
Foreign income $\equiv$ foreign expenditure.

Table 6.2 presents a descriptive SAM for China. An aggregated version of our estimated 1987 SAM for China is presented in Table 6.3.[3] The 4-sector version is used here for simplicity of exposition, although the CGE model for China is calibrated to the 10-sector version. In what follows, I will discuss how the data have been assembled and reconciled with the 1987 SAM for China, taking the commodity account and the activity account as the examples.[4]

---

[3] In the study of Rose *et al.* (1994) - see Section 7.5.1 for a further discussion, the 1990 I-O table is used. It has been estimated by proportionally extrapolating sectoral gross output and final demand of the 1987 I-O table based on actual GDP growth. This procedure will leave the I-O coefficients unchanged, thus making the estimated 1990 table not essentially different from the 1987 one. Such being the case, I preferred the 1987 I-O table. Besides, Chinese researchers think that the year 1990 is inappropriate as the base year because that year was during a period of economic retrenchment and is thus considered abnormal in terms of economic performance (Development Research Centre, 1993).

[4] See, for example, Pyatt and Round (1985) for a general discussion of the construction and uses of SAM.

Table 6.2   A descriptive social accounting matrix for China*

	Commodities 1a 1b 1c 1d	Activities 2a 2b 2c 2d	3	4a 4b	5	6	7	8	9	10
Commodities										
1a Agriculture		Intermediate demand		Household consumption		Government purchases	Fixed investment and change in stocks		Exports	Domestic sales
1b Industry										
1c Construction										
1d Services										
Activities										
2a Agriculture	Domestic supply							Export subsidies		Total sales
2b Industry										
2c Construction										
2d Services										
3 Value added		Net value added								Total net value added
4a Urban households			Labour income		Enterprise transfers to labour	Government transfers to households			Net foreign remittances	Total labour compensation
4b Rural households										
5 Enterprises			Capital income			Enterprise subsidies				Enterprise income
6 Government	Tariffs	Indirect taxes		Income taxes	Direct taxes and levies				Net foreign borrowing	Government income
7 Capital account				Household savings	Depreciation and enterprise savings	Government savings			Net foreign savings	Total savings
8 Foreign trade losses	Import subsidies					Foreign trade subsidies				
9 Rest of world	Imports									Foreign income
10 Total	Total absorption	Total costs	Total net value added	Total household expenditure	Enterprise expenditure	Government expenditure	Total investment		Foreign expenditure	

* Because of space limitation, account numbers only appear along the top margin, but readers should keep in mind that the accounts are in the same order in both rows and columns.

Table 6.3   *1987 social accounting matrix for China (100 million yuan)*[*]

	Commodities				Activities				3	4	5	6	7	8	9	10
	1a	1b	1c	1d	2a	2b	2c	2d								
Commodities																
1a Agriculture					688.5	1364.9	11.8	114.1				7.1	216.6			4588.4
1b Industry					614.5	6519.5	1556.5	1156.7				115.2	1486.3			13877.7
1c Construction					0	0	0	0				0	2430.6			2430.6
1d Services					170.7	1202.6	166.9	672.2				1207.6	137.7			4753.1
Activities																
2a Agriculture	4440.2													21.4	214.1	4675.7
2b Industry		12408.5												150.8	1253.7	13813
2c Construction			2430.6											0	0	2430.6
2d Services				4575.3										0	168.3	4743.6
3 Value added					3150.2	3837.6	622.5	2572.7								10183
4 Households									5600		1684.4	122.4			9.3	7416.1
5 Enterprises									4583			376.4				4959.4
6 Government	7.1	135.3	0	0	51.8	888.4	72.9	227.9			1171				68	2622.4
7 Capital account										1606.3	2104	621.5			-60.6	4271.2
8 Foreign trade losses												172.2				172.2
9 Rest of world	141.1	1333.9	0	177.8												1652.8
10 Total	4588.4	13877.7	2430.6	4753.1	4675.7	13813	2430.6	4743.6	10183	7416.1	4959.4	2622.4	4271.2	172.2	1652.8	

[*] Because of space limitation, account numbers only appear along the top margin, but readers should keep in mind that the accounts are in the same order in both rows and columns.

6.3.1 Commodity account

The row of the commodity account describes the domestic demand for goods and services. The column keeps track of absorption, which equals the value of goods and services supplied to the domestic market through domestic production and imports.

In the row, intermediate demand and government purchases have directly been taken from the input-output table (State Statistical Bureau, 1991). Household consumption and fixed investment and change in stocks have also been derived from the I-O table, but individual terms have been adjusted to eliminate the error item in final demand in the table.

In the column, the control figure for total imports and exports has been taken from the World Bank (1990c)[5] and the derivation is given in Table 6.4. Except for an 11-sector summary (State Statistical Bureau, 1991), there is one single column for net export in the 1987 I-O table. In order to incorporate product differentiation into our model for studying imperfect substitutability between imported and domestic goods, we reaggregated imports and exports in gross terms from the Customs Statistics (Customs General Administration, 1988) to fit the input-output table classifications. The control figure for total tariffs has been obtained from the State Statistical Bureau (1992a). Estimates for individual sectors have been made using rates of the *Official Customs Guide* (Customs General Administration, 1985). Estimates of the profits and losses on the imports and exports of foreign trade cooperations have been made by comparing the customs statistics figures (converted to Chinese yuan values) with the net exports at domestic prices, the losses of which are seen as import and export subsidies paid out of the government budget. Domestic supply (make matrix) has been calculated from the SAM itself as a residual.

6.3.2 Activity account

The row of the activity account describes the total sales of domestically-produced goods on both domestic and foreign markets. The column describes the total costs of production, including the demand for intermediate goods and factors of production and payments of indirect taxes.

Each entry in the row has already been described when explaining the column of the commodity account. In the column of the activity account, the control figure for total indirect taxes has been obtained from the World Bank (1990c). Indirect taxes for the industrial sectors have been taken from the State Statistical Bureau (1990b), while the corresponding figure for the services has been

---

[5] The control figure for any economic indicator in this chapter refers to aggregated number. By comparison with sectoral disaggregated figures, the control figure can easily be derived from official publications. Thus the figure can help to determine a sectoral figure that is difficult to obtain.

calculated using the rates of Wei *et al.* (1989). Enterprise subsidies are presented in Table 6.5. Dissagregated figures for the industrial sectors have been taken from the State Statistical Bureau (1990b). Net value added has been calculated as a residual.

Table 6.4   *Rest of world account control figures for 1987*

	Exports	Imports	Balance	Exports	Imports	Balance
		(billion US$)			(100 million yuan)	
1) Goods	39.437	39.629	-0.192	1468	1475	-7
2) Non-factor services	3.494	3.586	-0.092	130	133	-3
3) Factor services	1.027	1.191	-0.164	38	44	-6
4) Goods and services Total	43.958	44.406	-0.448	1636.1	1652.8	-16.77
5) Net foreign remittances			0.249			9.3
6) Net foreign borrowing						68
7) Net foreign saving (residual)						-60.6

Sources: World Bank (1990b,1990c); Customs General Administration (1988).

Table 6.5   *Government and foreign trade subsidies in 1987 (100 million yuan)*

1) Foreign trade subsidies[1]		172.23
2) Government subsidies		498.80
2a) Enterprise subsidies[2]	376.43	
2b) Consumer price subsidies[3]	122.37	
3) Total[4]		671.03

[1]   See the discussion about the column of the commodity account.

[2]   (3)-(1)-(2b).

[3]   State Statistical Bureau, 1990a.

[4]   294.60 (Price subsidies from the State Statistical Bureau (1992a)) + 376.43 (Subsidies for money-losing enterprises from the State Statistical Bureau (1992a)).

6.4   Structural parameters

Structural parameters for CGE models include behavioural parameters and calibrated parameters. These parameters need to be determined before any policy simulation by CGE models takes place.

6.4.1 Behavioural parameters

Behavioural parameters are a number of elasticities in the behavioural relationships of the model. In our model, this involves the elasticities in the CES production and import demand functions, CET export supply function, and ELES household demand function:

(a)   elasticity of substitution between value added aggregate and energy aggregate in sector i (σ_i);

(b)   income elasticity of consumption of consumer good i by household h (ζ_{hi});

(c)   price elasticity of substitution between imported and domestically-produced good i (ψ_i);

(d)   price elasticity of transformation between foreign and domestic sales of good i (η_i).

Table 6.6 shows the values chosen for these elasticities. They are *guess-estimated* and used to calibrate the model. In what follows, I will briefly explain these values.

Table 6.6   *Elasticity specification for production, demand and trade functions*

	σ_i	ζ_{hi}	ψ_i	η_i
Agriculture	0.3	0.7	0.6	0.9
Heavy industry	0.3	0.6	0.5	1.05
Light industry	0.3	1.0	0.7	0.3
Transport & Communication	0.3	1.1	-	0.9
Construction	0.3	0.9	-	-
Services	0.3	1.1	0.6	0.3
Coal	0.3	0.7	0.7	1.5
Oil	0.3	0.8	0.7	1.5
Natural gas[a]	0.3	0.8	-	-
Electricity[a]	0.3	0.9	-	-

[a] For natural gas and electricity, their imports and exports over time are set to be exogenous.

Sources: Dervis *et al.* (1982); Li *et al.* (1985); Shi (1991); Burniaux *et al.* (1992); own estimates.

The degree of substitutability between the value added aggregate and energy aggregate will affect the economic losses of energy scarcities and price increases. The higher the value of σ_i, the less expensive it is to decouple energy consumption from GDP growth when energy prices rise. In the absence of any empirical evidence, this parameter is assumed to be identical across sectors and is set at 0.3 following the assumption of Manne and Richels (1991a). That is, a 1% price increase will lead to a decline of 0.3% in the demand for energy. For the income elasticities of consumer demand, consistent with empirical evidence (Li *et al.*, 1985; Shi, 1991; Burniaux *et al.*, 1992), these are assumed to be higher for goods and services provided by the light industry, the transport and communication sector, and the service sector and lower for other sectors. With respect to the elasticities of substitution between domestic and imported goods, coal and oil, together with the nondurable goods from the light industry, are viewed as the homogeneous products and are thus assumed to be more substitutable in use than others. On the export side, in accordance with China's export pattern, it is assumed that the exports are generally more price elastic, particularly for the products from the energy sectors and the light industry.

6.4.2 Calibrated parameters

Calibrated parameters mainly refer to share and shift parameters. As is usual in CGE models, the values of these parameters are directly calculated from the model equilibrium conditions using the so-called calibration procedure (cf. Mansur and Whalley, 1984). In this section, I describe briefly the calibration procedure using the base year data and above-specified elasticity values.

The calibration procedure involves using the equilibrium conditions of the model and the benchmark year equilibrium data set to determine the share parameters, whose values are such that the equilibrium solution obtained to the model should reproduce the data observed in the base year (for example, the base year SAM) given the specification of numerical values of the behavioural parameters. Clearly, the calibration procedure also serves as a consistency check of the model equations, SAM data base, and calibrated parameters.[6]

6.4.2.1 Share and shift parameters in the value added aggregate
The share parameter is derived from minimizing the costs of capital and labour subject to the Cobb-Douglas aggregation. It is simply determined by the base year value share of these two factors 'and is assumed to remain constant over

---

[6] In programming our CGE model in GAMS, part of the computer code is written in order to make sure that the solution SAM is consistent with the base year SAM. This consistency test has been done by checking whether the row sum is equal to the column sum for each account. If they are not, some inconsistencies are present in the model. In this case, a close look at all the data sets and equation specification in the model is required.

time, with the exception of the agricultural sector in which the share parameter is expected to rise over time:[7]

$$\alpha_i = \frac{UK_i(t_o)K_i(t_o)}{UK_i(t_o)K_i(t_o)+W_i(t_o)L_i(t_o)}$$

Once the value of the share parameter is determined, the value of the shift parameter is given by

$$\bar{A}_i = \frac{VA_i(t_o)}{K_i(t_o)^{\alpha_i}L_i(t_o)^{1-\alpha_i}}$$

6.4.2.2 Share and shift parameters in the energy aggregate

In a similar manner, the values of the share and shift parameters in the energy aggregate can be determined as follows:

$$b_{ji} = \frac{a_{ji}}{\sum\limits_{j=7}^{10} a_{ji}} \qquad i = 1,...,10; \quad j = 7,...,10$$

$$\bar{B}_i = \frac{E_i(t_o)}{\prod\limits_{j=7}^{10} [VE_{ji}(t_o)]^{b_{ji}}} \qquad i = 1,...,10; \quad j = 7,...,10$$

6.4.2.3 Share and shift parameters in the import demand function

The share parameter can be derived from the optimal import demand function described in Section 5.7.1:

$$\mu_i = \left(\frac{PM_i(t_o)}{PD_i(t_o)}\right)\left(\frac{M_i(t_o)}{D_i(t_o)}\right)^{1/\psi_i}\left(1+\left(\frac{PM_i(t_o)}{PD_i(t_o)}\right)\left(\frac{M_i(t_o)}{D_i(t_o)}\right)^{1/\psi_i}\right)^{-1}$$

---

[7] This implies that future output growth in the agricultural sector will rely on increasing capital input.

The shift parameters are then calculated by

$$\Psi_i = Z_i(t_o) / \left(\mu_i M_i(t_o)^{-\varphi_i} + (1-\mu_i) D_i(t_o)^{-\varphi_i} \right)^{-1/\varphi_i}$$

6.4.2.4 Share and shift parameters in the export supply function
Similar to (c), the share and shift parameters in the export supply function are calculated as follows:

$$v_i = \left(\frac{PX_i(t_o)}{PD_i(t_o)} \right) \left(\frac{D_i(t_o)}{X_i(t_o)} \right)^{1/\eta_i} \left(1 + \left(\frac{PX_i(t_o)}{PD_i(t_o)} \right) \left(\frac{D_i(t_o)}{X_i(t_o)} \right)^{1/\eta_i} \right)^{-1}$$

$$\Phi_i = Q_i(t_o) / \left(v_i X_i(t_o)^{\phi_i} + (1-v_i) D_i(t_o)^{\phi_i} \right)^{1/\phi_i}$$

6.4.2.5 Marginal budget share and subsistence quantity parameters in the ELES household demand function
The computation of marginal budget share in the ELES function has already been described in Section 5.5.1. As for the subsistence quantity in the function, it is determined by[8]

---

[8] The ELES treats household demand as a function of its income and price. It is expressed by:

$$PC_i C_{hi} = PC_i \gamma_{hi} \overline{POP}_h + \beta_{hi} (YD_h - \sum_j PC_j \gamma_{hj} \overline{POP}_h) \qquad (i)$$

where the time index is dropped for simplicity.
Summing over the household demand for consumer goods, we have

$$\sum_i PC_i C_{hi} = (1 - \sum_i \beta_{hi}) \sum_i PC_i \gamma_{hi} \overline{POP}_h + \sum_i \beta_{hi} YD_h \qquad (ii)$$

Rearranging the equation (*ii*) gives

$$\sum_i PC_i \gamma_{hi} \overline{POP}_h = \frac{\sum_i PC_i C_{hi} - \sum_i \beta_{hi} YD_h}{1 - \sum_i \beta_{hi}} \qquad (iii)$$

Substituting equation (*iii*) into equation (*i*) and rearranging, we have the equation used for computing subsistence quantity in the ELES function.

$$\gamma_{hi} = \left(C_{hi}(t_o) - \frac{\beta_{hi}}{PC_i(t_o)} \left(\frac{YD_h(t_o) - \sum_j PC_j(t_o)C_{hj}(t_o)}{1 - \sum_j \beta_{hj}} \right) \right) / \overline{POP}_h(t_o)$$

6.5   Subsidiary data

Other than the SAM accounts and structural parameters, our CGE model requires, among other things, the following subsidiary data: capital composition matrix; sectoral depreciation rate and capital stock; sectoral allocation of investment; population; wage rate and labour supply; and factor-augmenting technical progress coefficients. This section documents the sources of these data used for this study.

For the 10-sector capital composition matrix used to convert investment in fixed assets by sector of destination into demand for capital goods by sector of origin, it has been aggregated in the following way from the semi-official 33-sector capital composition matrix, the latter being assembled by the State Planning Commission and the State Statistical Bureau.[9] First, I multiplied capital investment of the 33-sector of destination by the 33×33 capital composition matrix. This gave demand of each sector of destination for capital goods by sector of origin, which forms a 33×33 matrix. By summing over the corresponding demand for capital goods according to the sectoral classification discussed in Section 5.2, the 33×33 matrix was then aggregated into a 9×9 matrix in which there was one single aggregated sector for oil and natural gas. Next, I split the single aggregated sector into the oil sector and natural gas sector according to the intermediate demands of each sector for oil and natural gas in the input-output table (State Statistical Bureau, 1991). Finally, the 10-sector capital composition matrix was derived from dividing demand of each sector for capital goods by sector of origin by the corresponding capital investment by sector of destination. Table 6.7 shows the 10-sector capital composition matrix, the entries of which are assumed to remain unchanged over time.

The sectoral depreciation rate has been taken from Shi (1991) and the World Bank (1985b), but with a slight adjustment tailored to the needs of this study. Then sectoral capital stock in the base year has directly been calculated from dividing its value of depreciation by the corresponding depreciation rate, the former being derived from the input-output table (State Statistical Bureau, 1991). Its growth rates over time will be discussed in Section 7.2 when explaining key assumptions for the baseline scenario.

---

[9] I call the capital composition matrix a semi-official matrix because it has been assembled by the official institutes but has not formally been published. It is stored on disk only for internal use.

Table 6.7 *Capital composition matrix (10-sector by 10-sector)*

	AGRI	HIND	LIND	TRCO	CONS	SERV	CIND	OIND	NGAS	ELEC
AGRI	0.027259	0.023657	0.027489	0.033160	0.023740	0.023875	0.025865	0.024685	0.000000	0.031323
HIND	0.187920	0.357536	0.470762	0.324477	0.321468	0.127989	0.274599	0.279804	0.220177	0.341055
LIND	0.002807	0.002509	0.002830	0.003414	0.002444	0.002458	0.002663	0.002462	0.001292	0.003224
TRCO	0.006891	0.003602	0.006949	0.008382	0.006002	0.006034	0.006539	0.006176	0.001045	0.007918
CONS	0.759783	0.600912	0.476500	0.611906	0.632987	0.826212	0.675777	0.673458	0.769712	0.598849
SERV	0.015340	0.011783	0.015470	0.018660	0.013360	0.013432	0.014557	0.013415	0.007775	0.017631
CIND	0.000000	0.000000	0.000000	0.000000	0.000000	0.000000	0.000000	0.000000	0.000000	0.000000
OIND	0.000000	0.000000	0.000000	0.000000	0.000000	0.000000	0.000000	0.000000	0.000000	0.000000
NGAS	0.000000	0.000000	0.000000	0.000000	0.000000	0.000000	0.000000	0.000000	0.000000	0.000000
ELEC	0.000000	0.000000	0.000000	0.000000	0.000000	0.000000	0.000000	0.000000	0.000000	0.000000

Source: See text.

As far as the allocation of nominal fixed capital investment across sectors is concerned, the pattern of investment has for decades been characterized by a predominantly large share of investment in heavy industries. Consequently, the imbalance in the capital structure has led to a severe effect that can no longer be ignored. This pattern over time will thus be restructured to divert more resources to agriculture, to the existing bottleneck sectors such as energy, and to services, by comparison with the corresponding benchmark year shares in total nominal investment, which are determined through dividing the nominal fixed investment of sector of destination by total nominal investment. This not only fills the gap left from the past but also meets new demands.

The data on population and sectoral labour force in the base year have been derived from the State Statistical Bureau (1989, 1992a), while the projection for population and total labour force until the year 2010 will be discussed in Section 7.2. As for the sectoral wage rate in base year, it has directly been calculated from dividing sectoral wage bill by its labour force. In simulations, whether static or dynamic, however, the sectoral wage rate is treated as an endogenous variable that is assumed to be flexible enough to clear the labour market.

Total productivity growth rates over the period 1990-2010 have been taken from Yao *et al.* (1993), - see Section 7.2 for a further discussion.[10] Autonomous energy efficiency improvement for both fossil fuels and electricity over time is also discussed in that section.

6.6   Solution approaches

Four different approaches to finding numerical solutions to CGE models are identified in terms of solution procedure (see Robinson, 1989).

The first approach essentially amounts to specifying log-linear approximations of all equations of a CGE model and then solving the resulting linear equations for changes in the endogenous variables as functions of changes in the exogenous variables by inverting the resulting matrix of coefficients. This means that the solution describes the relative rates of change in the endogenous variables rather than the equilibrium levels of the endogenous variables (Bergman, 1990). The approach was first used in the first applied CGE model formulated by Johansen (1960) and has since been used in a number of applications. Dixon *et al.* (1982), for example, apply this technique in the ORANI model for Australia. The solution procedure has also been applied in the Swedish energy models (Bergman, 1980, 1982; Bergman and Lundgren, 1990). While enjoying the advantage of being a simple and relatively cheap method to apply, particularly for

---

[10] This implies that growth rates for real GNP are calculated endogenously. Alternatively, growth rates for real GNP are set to be exogenous, whereas the productivity growth rates become endogenous. The latter can be used to check whether the assumed productivity growth rates in the former case are reasonable.

large-scale models, the approach does have a limitation in flexibility of model specification inherent in being required to reduce the equations to a log-linear system. Moreover, the results are affected by approximation errors, and these errors tend to increase with the magnitude of changes in exogenous variables.[11]

The second approach involves treating a CGE model as just a collection of non-linear algebraic equations and solving them directly with a numerical solution technique. This technique is a special version of the Gauss-Seidel iteration procedure and does not require any evaluation of derivatives of excess demand equations: the algorithm simply adjusts the price in each sector in response to that sector's excess demand. The solution describes the equilibrium levels of the endogenous variables. Apart from the possible computational advantages by comparison with the first approach, the second one makes it possible to incorporate an explicit time dimension in the model. Thus the model is specified as a static within-period model linked by an intertemporal model that updates time-dependent variables. This is the technique first used by Adelman and Robinson (1978). It has later been used in applications to developing countries (cf. Dervis *et al.*, 1982).

The third approach is to express the problem as one of finding a fixed point in a mapping of prices to prices through excess demand equations and then solve it using an algorithm based on fixed-point theorem (cf. Scarf, 1984; Shoven and Whalley, 1992). Initially numerical solutions are determined by means of Scarf's algorithm (Scarf, 1967; Scarf and Hansen, 1973), but later faster variants of Scarf's algorithm, especially thanks to Merrill (1972) or Newton-type local linearization techniques, have become more commonly used (Bergman, 1990). A major advantage of this approach is that for models which satisfy the conditions of the fixed-point theorem, for example, homogeneous of degree zero in prices, the algorithm is guaranteed to find a solution. Like the second approach, the solution based on the third approach also describes the equilibrium levels of endogenous variables. A major disadvantage is that the algorithm gets very expensive to implement as the number of excess demand equations increases (Dervis *et al.*, 1982).

The last approach is to rework the economic specification so that it can be expressed as a maximization problem. The resulting system is then solved using a programming algorithm designed for constrained-maximization problems, and the solution yields shadow prices that can be interpreted as market prices. This approach is characterized as an extension of activity analysis and programming modelling and has the advantage that one can easily specify inequality constraints in the model. However, it may well not be at all convenient, analytically or

---

[11] See, for example, Dixon *et al.* (1982) for how to eliminate approximation errors arising from the linearization.

empirically, to express the model in terms of activity analysis (Dervis *et al.*, 1982).[12]

The CGE model described in Chapter 5 is highly non-linear, and has been implemented in GAMS (General Algebraic Modelling System), a widely distributed (non)linear programming package (Brooke *et al.*, 1988). The procedure involves two steps. The first one is to write the computer code. The code starts with declaring and defining all sets used, and declaring, defining and assigning all parameters, scalars and tables, followed by declaring and defining all variables and equations, and a mathematical description of all equations. The computer code ends with bounds and initial values of the variables, control commands, model statement, and output-generating statements. The phase of the process is independent of the solution algorithm. The second step is to solve the model numerically using the non-linear solver MINOS5.2. Given that the solver MINOS5.2 is based on a numerical solution algorithm, the solution approach adopted in our case belongs to the second type. It should be emphasized that MINOS5.2, being a non-linear programming package, can generally not find exact optimal solutions. Instead, it is guaranteed to find an approximate optimal solution or a local optimum (Brooke *et al.*, 1988).

6.7 Concluding remarks

In this chapter I have discussed some practical problems that arise when running the CGE model of the Chinese economy. These are related to data requirements, model calibration and solution approaches, which are crucial to the quality of the CGE model and its results as well as to the scope of its applicability. In what follows, some concluding remarks are given.

First, of all the issues considered, the work on constructing an SAM for China is the most difficult because of the shortage of data, particularly at a fairly disaggregated level. Although I managed to estimate the 10-sector SAM for China by assembling and reconciling the data from various sources, which involves some judgement on which sources are reliable and on how to modify certain data to make them consistent with the rest in the least arbitrary manner possible, the current version of the SAM should be considered preliminary, particularly those estimates of the actual values for sectoral imports and exports as well as their tariffs. This is partly because the 1987 I-O table only gives one single column for net export rather than imports and exports in gross terms. The second reason is that no official statistics but the *Official Customs Guide* provides the list of both minimum and general tariff rates by commodity. This has made it hard to know which rate was actually applied to what percentage of goods. The third reason is that the classifications in the *Official Customs Guide* are too

---

[12] See Ginsburgh and Wealbroeck (1981, 1984) for the activity analysis approach to CGE modelling.

disaggregated. This has made it very difficult to determine what average rate to use for an aggregated commodity like those used in the input-output table.

Second, because of data limitations, our CGE model, like most CGE models, has been calibrated rather than econometrically estimated. The implication of this treatment is that the results obtained through this model have a less sound basis than if the econometric approach that relies on empirical evidence was adopted. This suggests that the results reported in the next chapter should be interpreted with caution.

Third, although our model is flexible in formulating the demand side, the lack of a transformation matrix that defines the contribution of each producing sector to the composition of each of the final consumer goods and services (see Section 5.5.1 for a further discussion), in practice, limits its applicability to the case where there is only one single representative consumer and where the consumption category is the same as the sectoral classification of production. The implication is that the current version of this CGE model cannot be used to analyse the distributional effects of a carbon tax, although it is appropriate for calculating its economic costs. However, as discussed in Section 2.5.3, distributional aspects are considered to be important when designing a domestic carbon tax. Thus, it would be desirable to incorporate aspects of income distribution into the model if the required data become available.

7 Macroeconomic Analysis of CO_2 Emission Limits for China: A CGE Approach[1]

7.1 Introduction

In this chapter the time-recursive dynamic computable general equilibrium (CGE) model of the Chinese economy,[2] which has been described in Chapter 5 and calibrated in Chapter 6, is used to analyse the economy-wide impacts of alternative carbon limits through counterfactual simulations. In Section 7.2 the business-as-usual scenario is developed assuming no specific policy intervention to limit the growth rate of CO_2 emissions. In Section 7.3, counterfactual policy simulations are carried out to compute the macroeconomic and sectoral implications of two alternative carbon limits relative to the business-as-usual scenario, assuming that the carbon tax revenues are retained by the government. In Section 7.4, four carbon tax revenue recycling scenarios are constructed to illustrate the efficiency improvement from offsetting carbon tax revenues with reductions in indirect taxes relative to the carbon tax revenue retention scenarios above. A comparison with other studies for China in terms of both the baseline scenarios and the carbon constraint ones is presented in Section 7.5. The chapter ends with some concluding remarks.

7.2 The business-as-usual scenario

Before turning to macroeconomic analysis of CO_2 limits, we first have to develop the business-as-usual (BaU) scenario for economic development, energy consumption and CO_2 emissions in China, because any assessment of economic impacts of limiting CO_2 emissions starts with establishing a plausible baseline path. This BaU scenario assumes no policy intervention to limit the rate of CO_2 emissions, but does allow for anticipated changes in demographic, economic,

---

[1] This chapter forms the inputs to three papers in *Intereconomics* (Zhang, 1996a), *Journal of Policy Modeling* (Zhang, 1997b), and in *Economic Systems Research* (Zhang, 1997c).

[2] Recursive dynamic CGE models mean that a time sequence of single-period equilibria is computed for periods $t = 1, 2, \ldots$ Periods are related through the updating of some exogenous variables such as capital stock or demography (Gunning and Keyzer, 1995).

industrial and technological developments, and environmental policies not directly aimed at CO_2 emission reduction.

To develop the BaU scenario by using the time-recursive dynamic CGE model described in Chapter 5 involves two steps. The first step is to make a set of underlying baseline assumptions about how the exogenous variables in the model would evolve over the period till 2010. This involves updating time-dependent variables and revising certain parameters over time to reflect worldwide economic development and changes in tastes or technology. The second step is to use these assumptions to construct the BaU projections about the endogenous variables.

Table 7.1 shows what we assumed about some important exogenous variables in terms of the average annual growth rates of the variables over the period 1987-2010.

Table 7.1   *Some assumptions underlying the baseline scenario*
                 (average annual growth rates, %)

	2000/1987	2010/2000
GNP deflator	5.0	5.0
Total factor productivity	3.0	2.9
Population	1.37	0.88
Labour supply	1.43	0.53
Capital supply	9.4	8.5
World price of oil	2.8	3.0
World price of coal	2.2	2.3
World price of natural gas	2.8	3.0
World price of other imports	2.0	2.0
AEEI for fossil fuels[a]	1.0~4.5	1.0~2.7
AEEI for electricity[a]	0.0~0.3	0.0~0.2
Volume of government expenditures	8.0	7.0

[a] AEEI - Autonomous Energy Efficiency Improvement.

In the BaU scenario, the total productivity growth rates are set to be exogenous. This implies that the growth rates for real GNP are calculated endogenously. For this study, the total factor productivity growth rates during the period under consideration have been taken from projections by the Chinese Academy of Social Sciences (CASS) (Yao *et al.*, 1993), which are assumed to grow at an annual rate of 3% for the period 1987-2000 and 2.9% thereafter to 2010.

As discussed in Chapter 5, the GNP deflator serves as the price numéraire in the model. For simplicity, it is set at unity in the base year. For the period 1987-2010, the price numéraire is assumed to rise at an annual rate of 5%. This specification is inspired by the State Information Centre (1993) and reflects the fact that the domestic rate of inflation is higher than the world rate, which is assumed to be 2-3%.

Another important exogenous variable is population. It is assumed that China's basic long-term policy on family planning will continue to be implemented in the future. Thus, the population is expected to continue to rise, but with a declining growth rate. Following the projections by the CASS (Yao *et al.*, 1994), the current population of 1143 million in 1990 will rise to 1304.8 million in 2000 and to 1423.6 million in 2010. In the meantime, the labour force is assumed to grow from 527.8 million in 1987 to 634.5 million in 2000 and to 668.6 million in 2010 (Yao *et al.*, 1994).

The growth rates of capital supply are determined according to a Cobb-Douglas aggregation of capital and labour, using the average value of the share parameter in the value added aggregate discussed in Section 6.4.2. In this way, capital supply is calculated to grow at 9.4% per annum for the period 1987-2000 and 8.5% thereafter to 2010.

As far as the international energy prices are concerned, a rise in real energy prices is assumed to reflect depletion and increasing costs of exploration and mining. For the period 1990-2000, an average annual increase in oil prices in real terms is assumed to be 0.8%, and 1.0% for the period 2000-2010. Given the world oil price of US$ 22.3 per barrel in 1990, this implies a price of US$ 26.7 per barrel in 2010 at 1990 prices. This projections fall within the range of the specification by the Operating Agent for IEA-ETSAP/Annex IV (cf. Kram, 1993a).[3] With respect to internationally traded natural gas it is assumed that prices will follow the trend of oil prices. As to coal, world prices are assumed to rise less sharply than oil and natural gas prices. This favours the use of coal at the expense of oil and gas, raising the carbon-intensity and thus CO_2 emissions in the baseline scenario.

With respect to autonomous energy efficiency improvement (AEEI), its value has a crucial influence on the simulation results since *ceteris paribus* the higher the value the lower the growth of energy consumption and hence CO_2 emissions. For this study, the AEEI values vary among fuels and sectors and over time and

---

[3] The Energy Technology Systems Analysis Programme (ETSAP) was initiated in 1976 for the purpose of providing the International Energy Agency (IEA) with systems analysis capability to assist in establishing its priorities for research, development, and demonstration projects. Annex IV (Greenhouse Gases and National Energy Options) of ETSAP (1990-1993) was aimed at extending the application of national MARKAL models to the control of emissions of greenhouse gases. The Netherlands Energy Research Foundation (ECN), acting through the Policy Studies Department, was the Operating Agent for ETSAP/ Annex IV and Annex V (Energy Options for Sustainable Development). ECN is also serving as the Operating Agent for the ongoing ETSAP/Annex VI (Dealing with Uncertainty Together).

are set at $1.0 \sim 4.5\%$ per annum for fossil fuels and $0.0 \sim 0.3\%$ for electricity for the period 1987-2010. The lower AEEI values for the second period reflect the increasing difficulties in energy conservation, whereas the lower AEEI values for electricity reflect the difficulties in saving electricity and the increasing advantages of electricity use in new industrial processes. Moreover, compared with the values used in other studies, the higher AEEI value for China has been taken to reflect its potential for energy efficiency improvement and to track with the historical trends of such an improvement.

Other assumptions are related to the closure of the CGE model. The volume of government expenditures is assumed to grow at the same rate as the expected GNP. The government budget is assumed to be balanced over time, with total government revenues being equal to total government expenditures. Net foreign savings are assumed to be fixed in real terms at its benchmark-year value. No surplus or deficit on balance of trade is assumed to take place over time.

Table 7.2 summarizes the main macroeconomic results of the baseline simulation.[4] The baseline scenario is characterized by a rapid economic growth. As shown in Table 7.2, GNP is expected to grow at an average annual rate of 8.34% for the period from 1990 to 2000 and 7.55% thereafter to 2010. Although the calculated rates of GNP growth are lower than those achieved in the early 1980s and 1990s, they are well in line with the government targets of GNP growth rate, which are set at 8-9% per annum for the period 1990-2000 and at 7.2% thereafter to 2010.[5] Given that growth of the labour force is very small and declining significantly during the second period, which means that its contribution to output growth is also small in absolute terms and declining significantly in relative terms, such rapid economic growth is attributed partly to the increased factor productivity and mainly to increased capital stock. This reflects the long-established policy of the Chinese policymakers to rely on capital accumulation as the primary source of economic growth. But growth of capital depends on investment. Moreover, the existing bottlenecks in some sectors of capital-intensive nature (for example, energy and transport), and, more generally, imbalances in the output structure require adjustment in the capital structure.

---

[4] Because CGE models for resource allocation are unable to determine the absolute price level, it is meaningless to discuss a variety of price indexes under the baseline scenario, although their percentage deviations as a result of the imposition of a carbon tax relative to the baseline are important (see the next section). Thus, I only report changes in the underlying volumes.

[5] The Chinese Communist Party Central Committee's Proposals for National Economic and Social Development during the Ninth Five-year Plan period and up to the year 2010 set the goal of doubling China's 2000 GNP by the year 2010, although in principle it is subject to formal approval by the National People's Congress (the Chinese Parliament) in March 1996 (People's Daily (Overseas Edition), 29 September 1995).

Consequently, investment is expected to grow at a faster rate than GNP, as shown in Table 7.2. As can also be seen, the increasing exports form one of the driving forces behind China's booming economy.

Table 7.3 shows the energy-related results for the baseline scenario. Rapid economic growth will lead to increased energy consumption and hence CO_2 emissions, notwithstanding the reduced energy intensity of GNP.

Table 7.2 *Main macroeconomic results for the baseline scenario*
(average annual growth rates)

Volumes	2000/1987	2010/2000
GNP (%)	7.92[a]	7.55
Private consumption (%)	6.48	6.54
Investment (%)	8.81	7.81
Exports (%)	8.95	8.11
Imports (%)	6.86	5.12
Gross output (%)	8.25	7.86

[a] Converted to the period 1990-2000, this figure is equivalent to 8.34%, indicating the slowdown of economic growth during a period of economic retrenchment from 1988 to 1990.

Table 7.3 *Energy-related results for the baseline scenario*

	1990	2000	2010
Energy consumption (million tce)	987.0	1546.4	2560.4
Energy consumption per capita (tce)	0.86	1.19	1.80
Coal (million tons)	1055.2	1578.9	2418.2
Coal's share in total energy consumption (%)	76.2	72.9	67.5
Electricity (TWh)	623.0	1395.7	2745.2
Energy intensity of GNP (kgce/yuan)	0.717	0.504	0.403
Elasticity of energy consumption w.r.t. GNP[a,b]	0.56	0.55	0.68
Elasticity of electricity consumption w.r.t. GNP[a,b]	0.84	1.01	0.93
Average annual rate of energy conservation (%)[b]	3.6	3.46	2.21
CO_2 emissions (million tC)	586.9	898.9	1441.3
CO_2 emissions per capita (tC)	0.51	0.69	1.01

[a] w.r.t. is short for with respect to.
[b] The figures in 1990 are for the period 1980-1990, in 2000 for the period 1990-2000, and in 2010 for the period 2000-2010.

As shown in Table 7.3, total energy consumption is expected to rise from 987.0 Mtce in 1990 to 1546.4 Mtce in 2000 and to 2560.4 Mtce in 2010. Consequently, the baseline CO_2 emissions are expected to grow from 586.9 MtC in 1990 to 898.9 MtC in 2000 and to 1441.3 MtC in 2010 at an average annual rate of 4.4% for the period to 2000 and 4.8% thereafter to 2010. The slightly accelerated growth of CO_2 emissions during the second period is partly because economic growth, although somewhat slow in this period, still remains strong, and partly because of the reduced energy conservation rate as well as no significant change in the coal-dominated pattern of energy consumption. On a per capita basis, China's energy consumption of 0.86 tce in 1990 is expected to rise to 1.19 tce in 2000 and to 1.80 tce in 2010, whereas the corresponding CO_2 emissions of 0.5 tC in 1990 are expected to rise to 0.7 tC in 2000 and to 1.0 tC in 2010. Although the figures are doubled over twenty years, they are still well below the corresponding current world average levels, which were equal to 2.12 tce and 1.14 tC respectively in 1990.[6]

It should be emphasized that although the rapid growth of energy consumption and hence CO_2 emission per capita is attributed largely to rapid economic growth, it is attributed partly to the low population growth due to the implementation of a strict family planning policy.

7.3   Carbon abatement: counterfactual policy simulations

In this study six scenarios are considered.[7] First two scenarios as shown in Figure 7.1 are specified in terms of reductions in CO_2 emissions relative to the BaU path as follows:

Scenario 1:   20% cut in CO_2 emissions in 2000 and 2010 respectively;
Scenario 2:   30% cut in CO_2 emissions in 2000 and 2010 respectively.

Note that I use the BaU solution as the reference with which all alternative scenario solutions are compared, rather than the level of emissions in a single base year. The latter would be a much more restrictive target. It could be defended for industrialized countries which already enjoy a high level of output and consumption. It is also of policy relevance for industrialized countries because the Framework Convention on Climate Change commits industrialized countries to cut down emissions of CO_2 and other greenhouse gases to their 1990 levels by the year 2000 (Grubb and Koch et al., 1993). However, it is less defensible and relevant for developing countries which are still at an early stage of economic development (Blitzer et al., 1992).

In the two simulations the carbon tax revenues are retained by the government. In Sections 2.5 and 4.4, it has been argued that using the revenues raised to

---

[6] The world average levels of energy consumption and CO_2 emissions have been calculated based on data from Dean and Hoeller (1992) and British Petroleum (1993).

[7] The number of scenarios that can be developed with this CGE model is infinite. Given time constraints, however, only six scenarios are developed in the present study.

reduce a distortionary tax would lower the net adverse effects of carbon taxes by reducing inefficiency elsewhere in the economy. To determine how large the efficiency improvement might be, four tax reform simulations are constructed as follows:

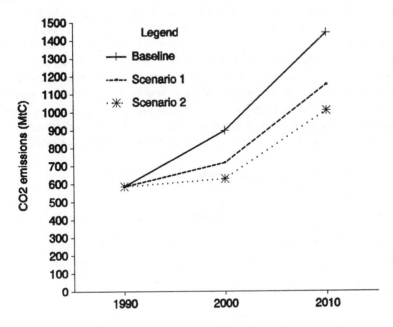

Figure 7.1 *CO₂ emissions in China under alternative scenarios*

Reforms 1a and 1b maintain the carbon tax of Scenario 1, but indirect tax rates for all sectors are equally reduced by 5% and 10% respectively.

Similarly, Reforms 2a and 2b maintain the carbon tax of Scenario 2, but indirect tax rates for all sectors are equally reduced by 5% and 10% respectively.

Such counterfactual simulations as shown in Table 7.4 allow us to compute the implications of two alternative carbon limits relative to the business-as-usual scenario, and the efficiency improvement of four indirect tax offset scenarios relative to the tax retention scenarios. In this section, I focus on discussing the first two simulations, whereas reporting the results from the four tax reform simulations will be the subject of Section 7.4.

7.3.1 Carbon taxes, fuel-specific tax rates and energy prices

Using the CGE model, the carbon tax required to achieve a 20% cut in CO_2 emissions in 2010 relative to the baseline is estimated to be 205 yuan per ton of carbon (tC). For Scenario 2, which is specified to achieve a 30% cut in CO_2 emissions in 2010, the carbon tax necessary is estimated to be 400 yuan/tC. The carbon taxes are estimated by trial and error using the GAMS 'SAVE' and

'RESTART' options through which a solution can be saved and reused (see the GAMS manual for details (Brooke *et al.*, 1988)). Such options are to preserve information that has been expensive to produce, and thus are always used by users of large models. More specifically, by using the 'SAVE' option first the baseline solution (that is, those at a zero carbon tax) is saved as work files. Then, an input file is written in which a carbon tax rate is specified, and which requests a separately restarted run of work files. The baseline solution is automatically used as a starting point for the carbon tax experiment. Moreover, work files can be used repeatedly with many input files containing different carbon taxes, until the carbon tax required to achieve a 20% cut in carbon emissions in 2010 relative to the baseline is found. A similar procedure is applied to the 30% cut scenario. Note that, unless specified otherwise, the carbon taxes estimated by our model are at current prices, that is, they are measured at the time of carbon emissions. Table 7.5 converts the carbon taxes into fuel-specific *ad valorem* tax rates.

Table 7.4   *Six policy simulations considered in this study*

Scenario	Changes in CO_2 emissions relative to the baseline	Changes in indirect tax rates relative to the baseline
Scenario 1	-20%	no change
Reform 1a	-20%	-5%
Reform 1b	-20%	-10%
Scenario 2	-30%	no change
Reform 2a	-30%	-5%
Reform 2b	-30%	-10%

Table 7.5   *Carbon taxes and fuel-specific tax rates in 2010*

	Scenario 1	Scenario 2
Carbon tax (yuan/tC)	205	400
Coal (%)	64.0	122.0
Oil (%)	14.4	28.2
Natural gas (%)	46.3	90.9
Electricity (%)	19.8	38.1

Two important observations can be made. First, the carbon taxes and the fuel-specific tax rates differ significantly among the two scenarios. As can be seen, a larger absolute cut in CO_2 emissions will require a higher carbon tax. A higher tax also implies higher fuel-specific tax rates because a carbon tax becomes higher relative to the baseline prices of fossil fuels. Moreover, carbon taxes and the fuel-specific tax rates rise at an increasing rate as the target of CO_2 emissions becomes more stringent, indicating that large reductions in carbon emissions can only be achieved by ever-larger increases in carbon taxes. Comparing Scenarios 1 and 2, for example, shows that the carbon tax goes up to 95% whereas the carbon reduction increases by only 50%. Second, although the same carbon tax is imposed in each scenario, tax rates differ considerably among different types of fossil fuels, depending on both the carbon content and the price of fuel in the absence of carbon taxes. Given that coal is the least expensive and gives rise to the highest CO_2 emissions per unit of energy content of all fossil fuels, it is not surprising that coal has the highest tax rates. A surprising result is that natural gas has a higher tax rate than oil,[8] although the former has fewer CO_2 emissions per unit of energy content. This is mainly because prices of natural gas in the absence of carbon taxes (or the pre-tax price in the baseline) are not rising faster than oil prices. As a result, tax rates for gas become higher relative to its prices,[9] although absolute levels of the tax per unit of energy content are higher for oil than for gas. As far as electricity is concerned, it is indirectly rather than directly affected by carbon taxes via the taxation of inputs used to generate electricity. This results in tax rates for electricity of about 20% in Scenario 1 and 38% in Scenario 2. They are considerably lower than those for coal, since coal accounts for only about 18% of total electric utility costs.

---

[8] The results are broadly consistent with the OECD's GREEN modelling results for North America, the Pacific and China (Burniaux *et al.*, 1991). Similar findings are also presented in the study of Ingham and Ulph (1991), who analysed the effect of carbon taxes on the UK manufacturing sector.

[9] This implies the increase in the post-tax price of natural gas relative to that of oil as shown in Table 7.6. This can be shown as follows. Denoting the pre-tax prices of natural gas and oil by P_g and P_o, the carbon tax per unit of carbon emitted by T^c, and the amount of carbon emitted per heat unit of natural gas and oil by e_g and e_o, the fractional change in the price of natural gas relative to that of oil as a result of the imposition of a carbon tax is given by

$$\frac{1 + (T^c e_g / P_g)}{1 + (T^c e_o / P_o)}$$

Rearranging, we have the change in the price of natural gas relative to that of oil

$$\frac{(P_g + T^c e_g) P_o}{(P_o + T^c e_o) P_g}$$

Given that the pre-tax price of natural gas is not rising faster than that of oil and that natural gas emits about quarter less carbon per heat unit than oil (see Table 3.15), the post-tax price of natural gas thus increases relative to that of oil.

Imposing fuel-specific tax rates contributes in turn to increases in prices of coal, oil, natural gas and electricity. Comparing the increases in Table 7.6 with the corresponding fuel-specific tax rates in Table 7.5, we can see that, as would be expected, they are almost the same.

Table 7.6 *Main macroeconomic effects for China in 2010*
(percentage deviations relative to the baseline; -: declines)

	Scenario 1	Scenario 2
GNP	-1.521	-2.763
Welfare[a]	-1.078	-1.753
Private consumption	-1.165	-2.972
Investment	-0.686	-1.832
Exports	-5.382	-7.447
Imports	-1.159	-2.128
Energy consumption	-19.468	-29.322
CO_2 emissions	-20.135	-30.112
Price elasticity of carbon abatement	-0.396	-0.317
Price of coal	64.954	123.095
Price of oil	15.296	29.144
Price of natural gas	46.813	90.564
Average price of fossil fuels	50.888	94.895
Price of electricity	22.785	43.256
Terms-of-trade	3.636	3.822
Nominal wage rate	-1.807	-3.043
Real exchange rate	-0.004	-0.021
User price of capital	-1.777	-4.228
Prices of exports	3.633	3.801
Prices of imports	-0.004	-0.021

[a] Measured in Hicksian equivalent variation as a percentage of the baseline household expenditures.

7.3.2 Macroeconomic effects

Curbing fossil fuel CO_2 emissions entails some corresponding reduction in energy consumption, and consequently, a decline in production itself. Basic economic theory can provide a rough order of magnitude estimates of this effect.

According to economic theory, the elasticity of output with respect to a factor should equal the factor's share in output.[10] Based on the data of 1987 SAM for China, which was discussed in Chapter 6, I calculate that the share of energy in China's GNP in the base year (1987) was 6.5%. That is, a 10% reduction in energy consumption will lead to a 0.65% decline in GNP. Table 7.6 shows that to achieve approximately 20% and 30% cuts in CO_2 emissions in 2010 requires about 19.5% and 29.3% cuts in energy consumption respectively. Accordingly, GNP in 2010 would fall by 1.27% (19.5% × 6.5%) and by 1.90% (29.3% × 6.5%) respectively, provided that the share of energy in China's GNP remains unchanged. Comparing this with Table 7.6, we can see that this order of magnitude estimate of the GNP effect is about what is predicted by this model, although the absolute values of the effect differ. Greater GNP losses predicted by this model imply that the share of energy in GNP tends to rise as the target of CO_2 emissions becomes more stringent. This reflects the assumed less-than-unitary elasticity of substitution between the inputs of energy, capital, and labour, such that energy has decreased less than GNP.

In addition to the GNP effect, Table 7.6 summarizes other main macroeconomic effects under the two scenarios. The results show that all the components of China's GNP and welfare in 2010 are negatively affected under the two CO_2 constraint scenarios compared with the baseline. Exports constitute

---

[10] This can be proved as follows. Suppose that total production Q is a function of capital K, labour L, and energy E. Let the price of output be P_q and the prices of capital, labour, and energy be P_k, P_l, P_e. Denoting the elasticity of output with respect to energy by ε_{qe} and the share of energy in output by s_e, then they can be defined as

$$\varepsilon_{qe} = \frac{\partial Q/\partial E}{Q/E} \qquad s_e = \frac{P_e E}{P_q Q}$$

According to the marginal product theory, the value of the marginal product of a factor should equal its price (Varian, 1993). Thus we have

$$P_q \frac{\partial Q}{\partial E} = P_e$$

Substituting $\partial Q/\partial E$ into ε_{qe} and rearranging, we have

$$\varepsilon_{qe} = \frac{\partial Q/\partial E}{Q/E} = \frac{P_e E}{P_q Q} = s_e$$

That is, the elasticity of output with respect to energy equals the share of energy in output. Similarly, we can prove that this conclusion also holds for capital and labour.

the final demand category that is reduced most.[11] Given the exogenous current account constraint, a decline in export volumes plus a rise in export prices also make import volumes decrease less than export volumes. Moreover, with the terms-of-trade improvement that tends to offset the deadweight losses arising from the imposition of carbon taxes, welfare, that is, the change in household real income that is measured in Hicksian equivalent variation, decreases less than GNP. As can also be seen, the level of carbon tax and the associated reductions in GNP and welfare rise as the carbon emission targets become more stringent. Moreover, it is indicated that the reductions in GNP and welfare tend to rise more sharply as the degree of the emission reduction increases. Put another way, the economic costs of incremental environmental policy actions increase with the level of emission reduction. This is reflected by, for instance, the increased elasticity of welfare with respect to emission reduction, which is 0.054 at a 20% required rate of reduction, and 0.058 at a 30% required rate of reduction. This is also reflected by the price elasticity of carbon abatement, which rises from -0.40 in Scenario 1 to -0.32 in Scenario 2. This increasing marginal cost of emission reduction implies that further reductions in CO_2 emissions are becoming increasingly more difficult. This finding also corresponds to other CGE studies.[12]

7.3.3 Effects on sectoral production and employment

Table 7.7 presents the percentage deviations of both aggregate and sectoral gross productions compared with the baseline. As can be seen, aggregate gross production tends to contract at an increasing rate as carbon dioxide emission targets become more stringent. However, changes in gross production vary significantly among sectors in both absolute and relative terms. The differing sectoral effects arising from the imposition of the carbon tax can be explained as follows. Meeting carbon emission targets via a carbon tax increases the prices of directly affected goods, such as coal, oil, and natural gas. As shown in Table 7.6, the more stringent the carbon emission targets, the more their prices increase. As prices rise, demand for the directly affected goods falls. A carbon tax also indirectly affects the prices of goods that utilize the targeted goods as factors in their production. For instance, Table 7.6 shows that the price of electricity in 2010 will rise by 23% in Scenario 1 and by 43% in Scenario 2 as an indirect effect brought about by increases in prices of fossil fuel input compared with the baseline, although carbon taxes are not directly imposed on electricity. This

---

[11] This is also observed in the CGE study of Glomsrød *et al.* (1992), who analysed the economic effects of stabilizing CO_2 emissions on the Norwegian economy.

[12] See, for example, Conrad and Schröder (1991) for Germany; Jorgenson and Wilcoxen (1993a, 1993b) for the United States; Beauséjour *et al.* (1995) for Canada; and Martins *et al.* (1993) for the global study.

indirect price effect exerts a further negative impact on gross production. Clearly, the combined direct and indirect effect will lead to a shift away from high-carbon energy, away from energy towards capital and labour, and away from carbon-intensive goods and services, although such shifts depend on the ability of producers and consumers to change to goods that are affected to a lesser extent by a carbon tax.

As shown in Table 7.6, the largest increase occurs in the price of coal in percentage terms as a result of the imposition of carbon taxes. It rises by 65% in Scenario 1 and by 123% in Scenario 2 relative to that of the baseline. In response to this price change, we expect that the coal sector is affected most severely in terms of the extent to which gross output falls under the two CO_2 constraint scenarios. This is confirmed in Table 7.7, which shows that gross production of the coal sector falls by as much as 26% in Scenario 1 and by 38% in Scenario 2. As would be expected, the substantial reduction in production growth will lead to a considerable decline in employment: the total number employed in the coal sector falls by 25% in Scenario 1 and by 36% in Scenario 2. Given the fact that about three-quarters of existing coal reserves are concentrated in the northern part, such negative impacts on the coal sector will pose serious regional implications.

Table 7.7 *Sectoral gross production in 2010*
(percentage deviations relative to the baseline; -: declines)

	Scenario 1	Scenario 2
Agriculture	-0.486	-0.281
Heavy industry	-2.463	-3.274
Light industry	-0.616	-0.416
Transport & Communication	-0.864	-14.146
Construction	-0.723	-1.444
Services	1.709	5.528
Coal	-26.498	-38.131
Oil	-2.072	-8.540
Natural gas	-20.781	-31.897
Electricity	-6.077	-10.722
Average - all sectors	-1.046	-1.900

The reduction in gross production in the natural gas sector is second largest.[13] This is a rather surprising result, since natural gas, with the lowest CO_2 emissions per unit of energy content, might be expected to benefit relatively from

---

[13] This is also observed in the study of Ingham and Ulph (1991), who analysed the effect of carbon taxes on the UK manufacturing sector.

the imposition of carbon taxes. The main explanation is that gas has the second highest tax rates, thus making the increase in the price of gas far larger than that of oil in percentage terms. This, as discussed earlier, can be attributed to the fact that prices of natural gas in the absence of carbon taxes are not rising very sharply. The considerable fall in production has a substantial negative effect on the employment in the gas sector, which is expected to fall by 20% in Scenario 1 and by 31% in Scenario 2.

Gross production also falls in the oil and electricity sectors. Because all of the four energy sectors are capital intensive, relatively large amounts of capital are released from these sectors. Given that the total amounts of capital available to the economy are fixed, the only way for all this additional capital to be absorbed in other sectors is for the relative price of capital to decrease. This explains why the user price of capital falls even faster than the wage rate in Scenario 2 as shown in Table 7.6.

In contrast to these negatively affected sectors, gross production increases are observed for the service sector. Moreover, the expansions rise at an increasing rate as carbon dioxide emission targets become more stringent. This is partly because the service sector utilizes a small proportion of intermediate inputs both directly and indirectly affected. The second reason is due to the output effect. As shown in Table 7.7, gross production falls in all sectors but the service sector. As a result, capital and labour are released from these sectors. Because factor supplies are fixed, the released amounts of capital and labour have to be absorbed in the service sector. Moreover, the more stringent the carbon dioxide emission targets, the more amounts of capital and labour have to be absorbed in the service sector and hence the more rapidly its production grows. The more rapid production growth in the sector relative to the baseline will lead to higher employment, which is expected to rise by 2% in Scenario 1 and by 6% in Scenario 2.

All in all, sectoral production and employment change much more than the aggregated macro variables. With the CO_2 constraints, the economy restructures towards labour-intensive sectors. This will come at the cost of lower GNP and welfare, provided that the tax revenues are retained by the government.

7.3.4 Effects on energy consumption and CO_2 emissions

Table 7.6 shows that to achieve approximately 20% and 30% cuts in CO_2 emissions in 2010 requires about 19.5% and 29.3% cuts in energy consumption respectively. As discussed in Section 5.8.4, the energy consumption reduction can be achieved through changes in both level and structure of economic activity, a change in energy input coefficients, and through a change in direct energy consumption by households. Table 7.8 clearly indicates the relative importance of each adjustment mechanism in terms of its contribution to energy consumption reduction in 2010. The results suggest that the overwhelming energy reduction is attributed to lower energy input coefficients under the CO_2 constraint scenarios compared with the baseline. But as the target of CO_2 emissions becomes more stringent, its role in reducing energy consumption becomes less because of the increased contribution by changes in both level and structure of economic activity. The contribution by direct energy consumption of households remains almost unchanged.

Table 7.8 *Breakdown of the contribution to energy consumption reduction in 2010 (%)*

	Scenario 1	Scenario 2
Due to change in aggregate production	4.69	5.66
Due to change in composition of aggregate production	9.80	13.76
Due to change in energy input coefficients	84.42	79.50
Due to change in direct energy consumption by households	1.10	1.09
Total change	100.00	100.00

Tables 7.9 and 7.10 show the percentage deviations of both sectoral energy consumption and CO_2 emissions as a result of the imposition of carbon taxes compared with the baseline. Four remarks can be made here.

First, as can be seen, energy consumption is reduced in all sectors. Accordingly, CO_2 emissions fall in all sectors. Looking at the rates of reduction for both sectoral energy consumption and CO_2 emissions, we can see that these are similar in both size and ranking across sectors. This implies that the amount of carbon emitted per unit of the sector's energy use remains largely unchanged.

Second, in relative (percentage) terms, energy consumption in the coal sector and the corresponding CO_2 emissions in 2010 are reduced most under both scenarios. This is because the largest increase in the price of coal leads to the largest decrease in the demand for it. In contrast to the largest effect on the coal sector, a slight reduction is observed for households. This is because the carbon taxes are applied only to industries for their use of fossil fuels, and not to households for their final energy demand.

Third, the reduction in total CO_2 emissions is larger than the reduction in total energy consumption. This is due to a shift in fuel consumption away from coal towards oil as shown in Table 7.11, the latter being less carbon-polluting than coal. Moreover, the larger the reduction in CO_2 emissions, the larger the extent to which such fuel switching takes place.

Fourth, the price elasticity of energy consumption rises as the carbon emission targets become more stringent. So does the price elasticity of carbon abatement because of the increasing marginal cost of emission reduction.

Table 7.9   *Sectoral energy consumption in 2010*
                 (percentage deviations relative to the baseline; -: declines)

	Scenario 1	Scenario 2
Agriculture	-0.486	-0.281
Heavy industry	-25.087	-35.827
Light industry	-26.163	-37.295
Transport & Communication	-15.983	-33.579
Construction	-15.011	-22.644
Services	-20.189	-27.402
Coal	-43.592	-59.028
Oil	-7.298	-16.565
Natural gas	-34.538	-50.330
Electricity	-21.205	-31.744
Households	-1.680	-2.504
Total[a]	-19.468	-29.322

[a] The corresponding price elasticity of energy consumption is -0.38 in Scenario 1 and -0.31 in Scenario 2.

Table 7.10   *Sectoral CO$_2$ emissions in 2010*
                  (percentage deviations relative to the baseline; -: declines)

	Scenario 1	Scenario 2
Agriculture	-0.486	-0.281
Heavy industry	-25.779	-36.729
Light industry	-26.703	-38.038
Transport & Communication	-17.277	-35.097
Construction	-16.230	-24.275
Services	-21.344	-29.033
Coal	-43.765	-59.237
Oil	-7.657	-17.019
Natural gas	-34.027	-49.657
Electricity	-21.606	-32.305
Households	-1.675	-2.497
Total	-20.135	-30.112

Table 7.12 shows the contribution by each fuel user to CO$_2$ emissions reduction in 2010. This depends on both the carbon intensity of each sector and the change in level of economic activity. The higher the carbon intensity of one sector and

the larger the reduction in its activity level, *ceteris paribus*, the bigger will be the contribution by that sector to CO$_2$ emissions reduction. The combined effects make, in absolute terms, the largest reductions occur in the heavy industry. As can be seen, almost half of the total reduction is realized in this sector. The contribution by the electricity sector ranks second, which is expected to be about 17% under both scenarios. In contrast to these large contributors, the contribution by households and agriculture is negligible, which together contribute to only about 1% of total reduction in CO$_2$ emissions.

Table 7.11   *Breakdown of fossil fuel use in 2010 (%)*

	Baseline	Scenario 1	Scenario 2
Coal	74.0	69.2	67.5
Oil	22.1	26.9	28.7
Natural gas	3.9	3.9	3.8
Total fossil fuel	100.0	100.0	100.0

Table 7.12   *Contribution by fuel user to CO$_2$ emissions reduction in 2010 (%)*

	Scenario 1	Scenario 2
Agriculture	0.023	0.009
Heavy industry	49.675	47.325
Light industry	4.613	4.394
Transport & Communication	9.164	12.448
Construction	2.531	2.531
Services	7.698	7.001
Coal	7.092	6.418
Oil	1.396	2.074
Natural gas	0.128	0.125
Electricity	16.581	16.578
Households	1.100	1.097
Total	100.000	100.000

7.4   Carbon tax revenue recycling scenarios

Imposing a carbon tax will raise government revenues. As Table 7.13 shows, a carbon tax of 205 yuan per ton of carbon would raise an additional government revenue of 261.3 billion Chinese yuan in 2010, whereas the corresponding amount of revenue for a carbon tax of 400 yuan/tC would be 448.5 billion Chinese yuan. Measured as a percentage of GNP in 2010, these government revenues correspond to 1.4% and 2.4% respectively. These amounts are certainly not negligible. How these revenues are used will affect the overall economic burden of carbon taxes. In this section, I will consider uses of these revenues raised to reduce the adverse effects of the carbon taxes discussed in the previous section by reducing indirect taxes. To determine how large the efficiency improvement might be by this option, four tax reform simulations are constructed.

Reforms 1a and 1b are based on Scenario 1. This means that the level of carbon tax in Reforms 1a and 1b is the same as in Scenario 1, which is 205 yuan per ton of carbon as shown in Table 7.14. But in the two simulations, part of the carbon tax revenues is recycled into the economy by means of equally reducing indirect taxes by 5% and 10% respectively.

Similarly, Reforms 2a and 2b are based on Scenario 2. They maintain the carbon tax of Scenario 2, but indirect tax rates for all sectors are equally reduced by 5% and 10% respectively.

Table 7.13   *Government revenues in 2010*

	Baseline	Scenario 1	Scenario 2
Cut in CO_2 emissions[1]	-	-20.1	-30.1
Government revenues[2]	42642.8	44569.0	45769.7
of which			
Indirect tax[2]	22525.7	22247.0	22043.7
Carbon tax[2]	-	2613.0	4485.5
Carbon tax revenues/GNP (%)	-	1.4	2.4
Change in government revenues[1]	-	4.5	7.3

[1] Percentage deviations relative to the baseline (-: declines).
[2] Measured in 100 million yuan.

Table 7.14 shows results in 2010 of the two simulations based on Scenario 1. Because there is a slightly smaller reduction in CO_2 emissions in Reforms 1a and 1b than in Scenario 1, the carbon tax revenues from Reforms 1a and 1b are slightly higher than Scenario 1. Moreover, as a result of the reduction in indirect tax rates, the total government revenues fall by 2.2% for Reform 1a and by 4.2% for Reform 1b relative to Scenario 1. The larger reduction in the revenues from Reform 1b is due to a larger reduction in indirect tax rates in Reform 1b than that in Reform 1a. With respect to the GNP effect, our results show that a 1.52% GNP loss under Scenario 1 is converted to a 1.51% loss in Reform 1a and a 1.47% loss in Reform 1b. These results suggest increased improvement in GNP if a larger reduction of indirect tax rates took place, although the improvement is small. The improvement is due to an increase in private consumption and international competitiveness of Chinese industries. By contrast, the welfare effect is markedly improved. As can be seen, a 1.08% loss under Scenario 1 is converted to a 0.41% loss in Reform 1a and even a 0.23% gain in Reform 1b.

Table 7.14   *Selected results for revenue experiments in 2010, related to Scenario 1*

	Scenario 1	Reform 1a	Reform 1b
Cut in CO$_2$ emissions[1]	-20.13	-20.06	-19.93
Carbon tax (yuan/tC)	205	205	205
Real GNP[1]	-1.52	-1.51	-1.47
Welfare[1,2]	-1.08	-0.41	0.23
Private consumption[1]	-1.17	-0.52	0.13
Exports[1]	-5.38	-5.19	-4.97
Government revenues[3]	44568.96	43576.31	42686.11
of which			
Indirect tax[3]	22247.03	21112.22	20075.2
Carbon tax[3]	2613.03	2616.56	2621.67
Change in government revenues[1]	4.52	2.19	0.10
Change in government revenues[4]		-2.23	-4.22

[1] Percentage deviations relative to the baseline (-: declines).
[2] Measured in Hicksian equivalent variation.
[3] Measured in 100 million yuan.
[4] Percentage deviations relative to Scenario 1 (-: declines).

Table 7.15 tells the same story, showing those results of the two simulations based on Scenario 2, although they are numerically different from the results for Reforms 1a and 1b. One important feature of the two simulations is the large improvement in GNP and welfare relative to Scenario 2, particularly for Reform 2b. This means that as the target of CO$_2$ emissions becomes more stringent, the positive effects of offsetting the carbon tax revenues with reductions in indirect taxes on GNP and welfare become more notable. This has been clearly shown in Figures 7.2 and 7.3, which illustrate the effects of reductions in indirect tax rates ranging from zero to 12% on GNP and welfare. As can be seen, a larger reduction of indirect taxes leads to a less negative economic growth and better welfare compared with the baseline. This finding has an important policy implication, as it suggests that if the target of CO$_2$ emissions becomes more stringent (that is, fossil fuels are taxed more heavily by carbon taxes), it will become more worthwhile to lower indirect taxes in order to reduce the adverse effects of a carbon tax.

Table 7.15   *Selected results for revenue experiments in 2010, related to Scenario 2*

	Scenario 2	Reform 2a	Reform 2b
Cut in CO_2 emissions[1]	-30.11	-30.01	-29.16
Carbon tax (yuan/tC)	400	400	400
Real GNP[1]	-2.76	-2.75	-2.18
Welfare[1,2]	-1.75	-1.09	-0.25
Private consumption[1]	-2.97	-2.33	-0.91
Exports[1]	-7.45	-7.23	-6.10
Government revenues[3]	45769.74	44791.17	44181.26
of which			
Indirect tax[3]	22043.67	20919.40	19877.03
Carbon tax[3]	4485.49	4493.53	4555.67
Change in government revenues[1]	7.33	5.04	3.61
Change in government revenues[4]		-2.14	-3.47

[1] Percentage deviations relative to the baseline (-: declines).
[2] Measured in Hicksian equivalent variation.
[3] Measured in 100 million yuan.
[4] Percentage deviations relative to Scenario 1 (-: declines).

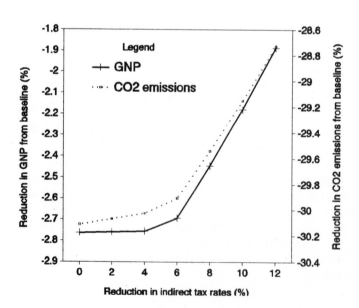

Figure 7.2   *GNP effect of indirect tax offset relative to Scenario 2*

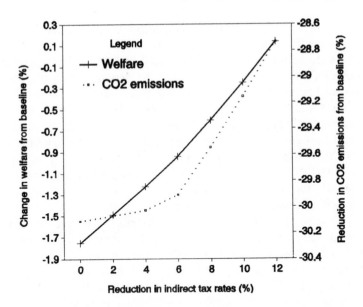

Figure 7.3   *Welfare effect of indirect tax offset relative to Scenario 2*

7.5   Comparison with other studies for China

Several studies have already addressed the issue of fossil fuel CO_2 emissions in China and/or of cost of their reductions. In this section, I will compare our results with those obtained by others, in terms of both the baseline scenarios and the carbon constraint scenarios.

Other studies considered for comparison include not only the well-known global studies based on GLOBAL 2100 and GREEN, but also the single-country studies by the US Argonne National Laboratory (ANL) and the Chinese Academy of Social Sciences (CASS). In Section 3.8, I already described these global studies. Thus prior to the comparison, I describe only briefly the two single-country studies. The ANL study is based on China's dynamic linear programming model, which has been adapted from the Dervis *et al.* (1982) model and aims to analyse the growth effect of various CO_2 mitigation strategies (Rose *et al.*, 1994). The CASS study, which is not a carbon abatement cost study, is based on China's system dynamics model and aims to forecast economic and social development in China under various circumstances (Yao *et al.*, 1994). Because these two single-country studies do not simulate the economic adjustment of a carbon tax, they are only referred to when comparing the baseline scenarios.

7.5.1   Comparison of the baseline scenarios across models

I start with comparing the baseline scenarios across models. Table 7.16 summarizes the results for the baseline scenarios across models.

As Table 7.16 shows, both GLOBAL 2100 and GREEN project very low growth rates of CO_2 emissions in China over the period 1990-2010. The results are as expected because both models set exogenously a very low growth rate of real GDP for China as specified by the Energy Modelling Forum (EMF) at Stanford University,[14] that is, at 4.25% per annum during this period.[15] Given that the target planned by the Chinese government is 8% to 9% annually over the period 1990-2000, which implies an annual growth rate of 3.9% to 4.4% over the period 1990-2010 even if zero rate of economic growth took place over the period 2000-10, the assumption about GDP growth in GLOBAL 2100 and GREEN is thus considered most unrealistic. Consequently, the baseline carbon emissions as estimated by GLOBAL 2100 and GREEN should be regarded as most optimistic from the point of view of CO_2 emissions. By contrast, the baseline CO_2 emissions estimated in the ANL study are the highest. This is also not surprising because energy conservation plays only a limited role in reducing CO_2 emissions in the ANL study, so that its baseline carbon emissions rise at almost the same rate as GDP. Given the optimistic economic growth and the most pessimistic view of energy conservation potential in the ANL study, the baseline emissions should be regarded as a worst-case scenario from the point view of CO_2 emissions.

In comparison with the two extremes above, the baseline CO_2 emissions estimated in the CASS study and by our model are in the middle of the pack. Comparing our results with the CASS results, however, we think that the latter are relatively optimistic in terms of both economic growth and energy conservation. In the CASS study, economic growth is estimated to be 8.8% annually over

---

[14] This is the so-called EMF-12 study. For models participating in EMF-12, including GLOBAL 2100 and GREEN, the common assumptions about, for example, GNP growth rate for each region have been adopted. For a complete description of EMF-12, see Gaskins and Weyant (1995).

[15] In the recent study based on MERGE (Model for Evaluating Regional and Global Effects of greenhouse gas reduction policies), Manne and Richels have updated their estimates by shifting from EMF-12 purchasing power parity concepts to measurements at market exchange rates. As would be expected, the level of 1990 GDP in China is much lower and at the same time the GDP growth rate is much higher than those in GLOBAL 2100. According to the preliminary results for the baseline obtained through MERGE, which are provided by Prof. Manne, China's GDP is expected to grow at an average annual growth rate of 7.94% over the period 1990-2010, and the corresponding baseline carbon emissions are expected to rise from 621 MtC in 1990 to 1384 MtC in 2010.

the period 1990-2010, which is on the high side of 8% to 9% planned for the period 1990-2000, and there is also a lack of backing from conventional wisdom for its energy conservation rate. In addition, the CASS study estimates the increased carbon intensity of energy use during the period under consideration. This contrasts sharply with the general finding that the carbon intensity of energy use is expected to be reduced, although the extent of such a reduction is not very large (see, for example, Zhang (1991), EWC/ANL/TU (1994) and this chapter).

From the preceding discussion, it thus follows that our estimates of the baseline CO$_2$ emissions represent the most plausible cases of all the models considered, although they are not a projection of what would actually happen in China if carbon emissions restriction was not imposed.

7.5.2 Comparison of the carbon constraint scenarios across models

Let us now compare the results for the carbon constraint scenarios. As mentioned earlier, because these two single-country studies do not simulate the economic adjustment of a carbon tax, a comparison of the carbon constraint scenarios only takes place between the global studies and our study.

Using GLOBAL 2100 benchmarked against the 1990 base year, Manne (1992) estimates, among other things, a GDP effect of 1% and 2% annual reductions in the rate of CO$_2$ emissions growth relative to the baseline. In absolute terms, these are equivalent to a 18.0% cut and a 32.7% cut in CO$_2$ emissions in 2010 relative to the baseline. Table 7.17 shows the results of the two simulations labelled as Scenarios 1% and 2% respectively. To allow the results to be compared more easily with our results, we interpolate the carbon taxes required and the associated GDP losses from achieving the same percentage of reductions in CO$_2$ emissions as those in our study, assuming that the change in carbon tax and the GDP effect are linear with respect to the magnitudes of reductions in carbon emissions. The results are given in parentheses in Table 7.17. This table also shows the simulation results of the 1% and 2% annual reductions in baseline carbon emission growth using GREEN as well as the required carbon taxes and the associated GDP losses converted to achieve the same percentage of carbon reductions as those in our study.

From Table 7.17, it can be seen that our estimates of a reduction in GNP growth are higher than those by GLOBAL 2100 and GREEN. Although it is difficult to provide a completely rigorous explanation for the differences between these results, which goes beyond the scope of this study, there are possibilities of identifying the sources of the differences, if not to quantify their significance.

First, the economic costs of a carbon constraint are determined to a large extent by the baseline of CO$_2$ emissions. The more rapid the growth of uncontrolled emissions under business-as-usual, the larger the size of the gap between uncontrolled emissions and a particular target and hence the higher the costs to meet the target. Given that our baseline of carbon emissions is higher than that in GLOBAL 2100 and GREEN, it should thus come as no surprise that our estimates of GNP loss are higher than those by GLOBAL 2100 and GREEN.

Table 7.16 *A comparison of the baseline scenarios across models in 2010*

	Our CGE model	GLOBAL 2100	GREEN	ANL	CASS
CO_2 emissions (MtC)[a]	1441.3	937.0	1363.0	1959.0	1427.0
Growth rate of CO_2 emissions (%)[b]	4.59	1.92	4.12	7.73	5.07
Growth rate of GNP (%)[b]	7.95	4.25	4.25	8.00	8.80
Energy conservation rate (%)[b]	2.84				3.72
Change rate of CO_2 emissions per unit of energy use (%)[b,c]	-0.28				0.32

[a] CO_2 emissions in the ANL are measured in tons of carbon dioxide. Divide by 3.67 to convert to tons of carbon.
[b] Average annual rate over the period 1990-2010.
[c] -: declines.

Sources: Manne (1992); Martins *et al.* (1993); Rose *et al.* (1994); Yao *et al.* (1994); own calculations.

Table 7.17 *A comparison of CO$_2$ emission reductions, carbon taxes and growth effect across models in 2010*

	CO$_2$ emissions[a]	Carbon tax[b]	GNP (GDP)[a]
GLOBAL 2100[c]			
Scenario 1%	-18.036	57.999	-0.783
Scenario 2%	-32.657	165.837	-2.127
Scenario 1	(-20.135)	(73.480)	(-0.976)
Scenario 2	(-30.112)	(147.066)	(-1.893)
GREEN[c]			
Scenario 1%	-17.535	8.000	-0.200
Scenario 2%	-32.135	20.000	-0.500
Scenario 1	(-20.135)	(10.137)	(-0.253)
Scenario 2	(-30.112)	(18.337)	(-0.458)
Our CGE model			
Scenario 1	-20.135	17.929	-1.521
Scenario 2	-30.112	34.983	-2.763

[a] Percentage deviations relative to the corresponding baseline (-: declines).
[b] Measured in US dollars per ton of carbon. In GLOBAL, carbon taxes are measured at 1990 prices, in GREEN at 1985 prices, and in our model at 1987 prices.
[c] The figures in parentheses result from interpolating the carbon taxes required and the associated GDP losses that have originally been estimated by GLOBAL 2100 and GREEN in order to achieve the same percentage of carbon reductions as those in our study.

Sources: Manne (1992); Martins *et al.* (1993); own calculations.

The second reason is related to the sectoral aggregation of the Chinese economy. Our model is much disaggregated compared with GLOBAL 2100, the latter including a macroeconomic growth model with only one final output good in its highly aggregated representation of the economy. Our model is also relatively disaggregated compared with GREEN, the production block of which includes only eight sectors. This implies less substitutability in our model and hence leads to higher economic costs, since the implicit assumption of aggregation is that all output and resources within one aggregate are perfect substitutes.

The third reason is related to the model type. Our model is a single-country CGE model. It is able to calculate the economic effects of unilateral carbon taxes on China, assuming that its trade partners do not react to carbon taxes. By contrast, GREEN is a global CGE model, with China being treated as a separate region. This global model allows for computing regional economic implications of region-specific carbon taxes under the assumption that all trade partners do react to carbon taxes, although the extent of reaction may vary significantly among partners. Because the global model takes into account actions of all trade partners, this may make its results different from those of a single-country model. For instance, although both GREEN and our model estimate that the coal sector in China is affected most severely as a result of the imposition of carbon taxes, their estimates of the effect on energy-intensive industry quite differ. In our model, the energy-intensive industry (the heavy industry) in China is estimated to be negatively affected by the imposition of unilateral carbon taxes because this sector uses a large proportion of intermediate inputs both directly and indirectly, while in GREEN, its output is virtually unchanged (Burniaux *et al.*, 1991) because of the relative improvement in Chinese energy intensive goods' competitiveness via trade reallocation. The differing effects brought about by the imposition of unilateral carbon taxes or by the imposition of regional carbon taxes may partly explain why the estimates of China's GNP loss by our model are higher than those by GREEN. This suggests that the economic cost of carbon abatement in China would not appear to be that high if actions of its trade partners were taken into account. This finding is also consistent with results from studies based on the game theory, which demonstrate that cooperative outcomes are better than noncooperative ones in terms of the cost-effectiveness of emission reduction (cf. Barrett, 1990).

Let us now explain the differences between the carbon taxes required across models.

First, the magnitude of the carbon tax depends on the size of the gap between uncontrolled emissions and a particular target. The larger the gap, the more carbon is to be reduced to meet the emission target and hence the higher will be the carbon tax required. This explains why our estimates of carbon tax are higher than those by GREEN and why a larger absolute cut in CO_2 emissions will require a higher carbon tax, as shown by the estimates with GLOBAL 2100 and our model. When explaining why GLOBAL 2100 gives higher carbon taxes than our model, however, this explanation does not hold.[16] We will come back to this issue later.

Another reason why the carbon tax is lower in GREEN than that in our model is related to the benchmark value of domestic energy prices in China. In GREEN, although the year 1985 is chosen as the base year, China's input-output table for 1981 is used, while in our model China's 1987 input-output table is used. Given that fossil fuels, particularly coal, were more heavily subsidized in

---

[16] This clearly indicates that model structure does matter in comparing model outcomes. Our model and GREEN are similar in the sense that both models are of CGE type and thus allow for the straightforward comparisons. By contrast, GLOBAL 2100 is an optimization model with a detailed treatment of the energy sector but a highly aggregated description of the economy (see Section 4.3).

1981 than in 1987,[17] it is not surprising that GREEN requires lower carbon taxes than our model, because GREEN has lower baseline prices of fossil fuels.

Let us return to the carbon tax level of GLOBAL 2100. There have been a number of critiques of the study of Manne and Richels (1990). These critiques include that of Williams (1990), Hogan (1990) and Morris *et al.* (1991), who considered that the carbon tax level estimated by GLOBAL 2100 is too high. Here I mention two factors that are of crucial importance for the high carbon tax level of GLOBAL 2100.[18]

The first factor is the autonomous energy efficiency improvement (AEEI). This parameter is considered to be crucial to limiting the tax level required. In a series of studies based on GLOBAL 2100 (Manne and Richels, 1990, 1991a, 1991b; Manne, 1992), however, Manne and Richels have been unduly pessimistic in choosing the values of the AEEI parameter. For instance, Manne (1992) assumes that for China the AEEI is 1% per year. This value is much lower than 3.6% observed over the period 1980-90 (see Section 3.5). The assumed low value of the AEEI parameter makes energy conservation as modelled in GLOBAL primarily a result of the imposition of a carbon tax. This in turn would lead to a high level of carbon taxes in order to reduce CO_2 emissions to the target level.

Second, the abatement options considered and the estimated costs of the options will also affect the carbon tax rates required. As discussed in Section 4.3, the core of GLOBAL 2100 is the ETA module with the explicit, process-oriented description of energy supply technologies. Williams (1990) thinks that the options for reducing CO_2 emissions are much broader than those considered by Manne and Richels (1990). Moreover, Williams argues that ETA may overstate the costs of important alternative low carbon-polluting energy technologies because no account is taken of near-term opportunities for cost reduction for these options. Where there are few economically feasible substitutes available, the effectiveness of a carbon tax is likely to be much more limited. Thus, in order to make these alternative technologies become competitive with traditional high carbon-polluting technologies, a higher carbon tax is required than would otherwise have been the case.

---

[17] As shown in Table 3.12, the ratio of domestic coal price to world coal price was 0.84 in 1989, while the corresponding figure is only 0.45 in GREEN (Lee *et al.*, 1994). If we define fossil fuel subsidies as the difference between domestic fossil fuel prices and their world prices, this means that coal is more heavily subsidized in GREEN than in our model.

[18] For detailed critiques, see Williams (1990) and Hogan (1990).

Despite numerical differences across models in the carbon tax rates and their associated costs, the following consensuses emerge:[19]

First, a larger absolute cut in CO_2 emissions will require a higher carbon tax. Moreover, carbon tax rises at an increasing rate as the target of CO_2 emissions becomes more stringent, indicating that large reductions in carbon emissions can only be achieved by ever-larger increases in carbon taxes.

Second, the associated GNP losses rise as the carbon emission targets become more stringent. Moreover, they tend to rise more sharply as the degree of the emission reduction increases.

Third, China would be one of the regions hardest hit by carbon limits. This is reflected by the fact that China's GNP losses under less restrictive carbon limits are in the same range as the often reported estimates for industrialized countries under very restrictive carbon limits.

Note that the preceding discussion focuses on China. Nevertheless it is worthwhile comparing the magnitude of carbon tax across regions because it forms a basis for China's development of joint implementation projects with other countries. Using the same labels as those in Table 7.17, Table 7.18 shows the carbon taxes across regions of the 1% and 2% annual reductions in baseline carbon emission growth using GREEN, as well as the required carbon taxes converted to achieve the same percentage of carbon reductions as those in our study, with the latter given in parentheses.

Table 7.18   *Carbon taxes across regions in 2010 (in 1985 $ per ton of carbon)*[a]

	USA	Japan	EEC	Total OECD	China	World
Scenario 1%	39	46	71	48	8	34
Scenario 2%	139	116	180	152	20	105
Scenario 1	(53.4)	(55.9)	(85.7)	(62.7)	(10.1)	(45.1)
Scenario 2	(120.3)	(103.1)	(158.6)	(132.3)	(18.3)	(92.9)

[a] The figures in parentheses result from interpolating the carbon taxes required and the associated GDP losses that have originally been estimated by GREEN in order to achieve the same percentage of carbon reductions as those in our study.

Sources: Martins *et al.* (1993); own calculations.

---

[19] The first two consensuses are also in line with general findings from other CGE studies; see e.g. Conrad and Schröder (1991) for Germany; Jorgenson and Wilcoxen (1993a, 1993b) for the United States; Beauséjour *et al.* (1995) for Canada; and Martins *et al.* (1993) for the global study.

Comparing the carbon tax levels in Table 7.18 with those in Table 7.17, we can see that there are significant differences in the carbon taxes required across countries and regions in order to achieve the same percentage of emission reductions relative to the baseline. This points to opportunities for international trade in carbon emission permits to reduce the global costs of abating CO$_2$ emissions. However, it is unlikely that a global regime of tradeable carbon permits will emerge in the near future. Thus, as a preliminary step towards that regime, joint implementation (JI) mechanism, although not without conceptual and operational problems (see Section 2.5.6), should be considered a means of abating global CO$_2$ emissions effectively.

Then, between which parties should JI take place? As shown in Table 7.18, the carbon taxes are much higher in the industrialized countries than in the developing countries. This is, among other things, due to their already relatively energy-efficient economies, their limited possibilities for substituting less polluting energy sources and their already high pre-carbon tax energy prices as a result of existing energy taxes. But their differences are far less than that between the industrialized countries and developing countries. Then the question arising from this is whether such differences are large enough to justify every JI deal between the industrialized countries, not least due to the assumed transaction costs.[20] But Table 7.18 clearly indicates that there is large potential for JI deals between the industrialized countries and developing countries. In addition to its cost-effectiveness, there are other arguments in favour of such deals. For example, in the developing countries, there is a pressing need for reform of their energy sectors, on both environmental and economic grounds. Thus, there is a widespread need for transfers of financial resources, technology and expertise from the industrialized countries. Such transfers may be encouraged by JI. JI projects will also contribute towards reducing local environmental problems, which will benefit both the industrialized countries and developing countries. For example, Japan is extremely concerned about cross-border pollution in the form of acid rain originating from coal-fired power plants on the eastern coast of China. Clearly, JI projects for increased energy efficiency and fuel switch can make a positive contribution to this kind of problems. This also applies to the acid rain problem between the Nordic countries and their neighbouring countries.

---

[20] For instance, the transaction costs associated with the two Norwegian pilot JI projects in Poland and Mexico have been estimated at about 10% of the budget for the two projects (cf. Barrett, 1995). Of course, the first two JI projects are the demonstration projects, and we should expect transaction costs to fall once generic procedures have been established.

7.6   Concluding remarks

As a starting point of macroeconomic analysis of carbon emission limits, a baseline scenario has first been developed under a set of assumptions about the exogenous variables. The calculation results show that a rapid growth of the Chinese economy will take place until the year 2010. Consequently, this will lead to increased energy consumption and hence CO_2 emissions, despite substantial energy efficiency improvement. Moreover, a comparison with other studies for China has shown that of all the models considered, our estimates of the baseline CO_2 emissions represent the most plausible cases from the point view of CO_2 emissions.

Then, using a time-recursive dynamic CGE model and assuming that carbon tax revenues are retained by the government, Section 7.3 analyses the implications of two less restrictive scenarios under which China's CO_2 emissions in 2010 will be cut by 20% and 30% respectively relative to the baseline. Our main findings can be summarized as follows.

First, a larger absolute cut in CO_2 emissions will require a higher carbon tax. Higher tax also implies higher prices of fossil fuels. Moreover, carbon tax rises at an increasing rate as the target of CO_2 emissions becomes more stringent, indicating that large reductions in carbon emissions can only be achieved by ever-larger increases in carbon taxes and hence prices of fossil fuels.

Second, even under the two less restrictive carbon emission scenarios, China's GNP drops by 1.5% and 2.8% and its welfare measured in Hicksian equivalent variation drops by 1.1% and 1.8% respectively in 2010 relative to the baseline, indicating that the associated GNP and welfare losses tend to rise more sharply as the degree of the emission reduction increases. Given the often reported losses of 1-3 per cent of GDP in industrialized countries under very restrictive carbon limits, the results also support the general finding from global studies that China would be one of the regions hardest hit by carbon limits.

Third, although aggregate gross production tends to decrease at an increasing rate as the carbon dioxide emission target becomes more stringent, changes in gross production vary significantly among sectors in both absolute and relative terms. Of the ten sectors considered, we found that the coal sector is affected most severely in terms of output losses under the two CO_2 constraint scenarios. Consequently, this will lead to a considerable decline in the sector's employment. This, combined with the fact that about three-quarters of existing coal reserves are concentrated in the northern part, suggests that special attention should be paid to the sectoral and regional implications when designing a domestic carbon tax.

Fourth, of the four adjustment mechanisms considered, lower energy input coefficients contribute to the bulk of energy reduction and hence CO_2 emission reduction in 2010 under the two CO_2 constraint scenarios, followed by a change in the structure of economic activity. With respect to the contributions to CO_2 abatement in 2010, although in relative terms energy consumption in the coal sector and the corresponding CO_2 emissions in 2010 are reduced most under both scenarios, in absolute terms, the largest reductions occur in the heavy industry.

In Section 7.4, we have computed the efficiency improvement of four indirect tax offset scenarios relative to the two tax retention scenarios above. The four simulations labelled as Reforms 1a, 1b, 2a and 2b respectively show that if these revenues were used to offset reductions in indirect taxes, the negative impacts of carbon taxes on GNP and welfare would be reduced. Moreover, as shown by Reforms 2a and 2b as well as in Figures 7.2 and 7.3, the efficiency improvement tends to rise as the target of CO_2 emissions becomes more stringent (that is, fossil fuels are taxed more heavily by carbon taxes). This suggests that it would become more worthwhile to lower indirect taxes in order to reduce the adverse effects of a carbon tax.

In Section 7.5, a comparison with other studies for China has been made. It has been indicated that our estimates of the reduction in GNP growth are higher than those by GLOBAL 2100 and GREEN in order to achieve the same percentage of reductions in CO_2 emissions relative to the baseline. This might be related to three factors. First, our baseline of carbon emissions is higher than that in GLOBAL 2100 and GREEN. Second, our model is relatively disaggregated compared with both GLOBAL 2100 and GREEN. This implies less substitutability in our model, leading to higher economic costs. Third, model types matter. While in our single-country model one branch of industry is estimated to be negatively affected under the carbon constraints, this would not always be the case in a global model such as GREEN because of the relative improvement in Chinese branch goods' competitiveness via trade reallocation. The differing effects brought about by the imposition of unilateral carbon taxes or regional carbon taxes could be part of the explanation for the higher GNP losses in our model.

With respect to the carbon taxes required to achieve the same percentage of carbon reductions in 2010 relative to the baseline, our estimates are on the one hand higher than those by GREEN. This is because GREEN has a smaller size of the gap between the uncontrolled emissions and the emission target, and because GREEN has lower baseline prices of fossil fuels. On the other hand, our estimates are lower than those by GLOBAL 2100. This is because GLOBAL 2100 assumes lower values of the AEEI parameter, and because GLOBAL 2100 considers limited options for reducing CO_2 emissions and overstates the costs of some important alternative low carbon-polluting energy technologies.

Finally, comparing carbon tax levels across the regions considered shows that the carbon taxes required in China in order to achieve the same percentage of emission reductions relative to the baseline are much lower than those for both the industrialized countries and the world average. This suggests that the JI mechanism as a preliminary step towards a global regime of tradeable carbon permits should be considered a means of reducing global CO_2 emissions effectively. This mechanism will not only help China, which is becoming an important source of future CO_2 emissions, alleviate the suffering from possible future carbon limits, but also act to lower the costs of undertaking carbon abatement in the industrialized countries that are currently responsible for the majority of global CO_2 emissions and hence to reduce the competitive disadvantage and carbon leakage associated with purely unilateral policies in these countries. Worldwide, this will achieve global carbon abatement at a lower overall cost than would otherwise have been the case.

8 Cost-Effective Analysis of Carbon Abatement Options in China's Electricity Sector[1]

8.1 Introduction

Faced with the likely threat of global warming, society is being asked to take necessary measures in response. In Chapter 2, it has been pointed out that a well-designed limitation strategy is of great policy relevance. Thus, it seems logical, once the preventive position is adopted, to assume that policymakers would wish to pursue such a strategy in the most cost-effective manner in order to obtain political support.

This chapter attempts to identify the most cost-effective options for carbon abatement in the electricity sector and is thus devoted to satisfying energy planning requirements in the CO_2 context. There were at least four reasons for focusing on this sector. First, as discussed in Chapter 3, the electricity sector is a major consumer of primary energy. Second, there are many more interfuel substitution possibilities in the electricity sector compared with other sectors. The third reason concerns joint implementation projects. Section 2.5.6 argues that projects of these kinds should be related to development priorities of developing countries. As discussed in Section 3.7, projects for both increased generation efficiency and interfuel substitution in the electricity sector are clearly the priorities to be taken in China. Thus, the results obtained from studying the sector can contribute to forming a necessary basis for China's development of joint implementation projects with other countries. Fourth, although it would have been desirable, time constraints did not allow studying the Chinese energy system as a whole.

This chapter proceeds as follows. Section 8.2 presents a brief introduction of two types of models commonly used to derive fuel choice in electricity generation. In Section 8.3, I describe the MARKAL model in detail, a generalized model for the least-cost energy planning, from which a technology-oriented optimization model for power system expansion planning is developed for the present study. Section 8.4 presents all the alternative electricity generation technologies considered in this study. The description of these technologies by their technical, economic and environmental parameters is the subject of Section 8.5. In Section 8.6, a comparison of alternative power plants is made in terms of

---

[1] A large part of this chapter forms the basis for an invited paper for a special issue of the *Energy Sources* on the theme 'Energy, Environment and Sustainable Development' (Zhang, 1997a).

both generation costs and marginal costs of carbon abatement. In Section 8.7, the business-as-usual scenario is developed for electricity supply. In Section 8.8, the CO_2 constraint scenarios are examined with respect to the baseline one. The chapter ends with some concluding remarks.

8.2　Fuel choice: econometric approach versus optimization approach

Two types of models are commonly used to derive fuel choice in electricity generation for an exogenously-determined electricity demand: one is based on econometric approaches, and the other is to use optimization models. The econometric approach relies largely on econometric estimates of behavioural and interfuel substitution of power supply in response to changes in the relative fuel prices. The variables concerned, for example, flows of economic activities and energy flows, are treated in monetary terms. By contrast, the optimization approach is technology-oriented, thus describing energy flows of electricity generation in physical terms and explicitly representing alternative electricity generation technologies, including newly-emerging power supply technologies.

8.2.1　Econometric approach

As discussed in Chapter 5, within the computable general equilibrium framework, production or cost functions of Leontief, Cobb-Douglas, and/or CES (constant elasticity of substitution) forms are most commonly used for operational simplicity. These function forms, however, impose a prior restriction on the Allen elasticities of substitution:[2] the CES cost function assumes a constant elasticity of substitution between a pair of inputs, the Cobb-Douglas function imposes a unitary elasticity of substitution, and Leontief function implies zero elasticity of substitution between factor inputs. Moreover, none of these functions is flexible enough to allow differences in the elasticities of substitution between pairs of inputs when more than two exist.[3]

Our attention will now be restricted to fuel choice in electricity generation. This allows us to employ a more flexible specification capable of exploring substitution effects caused by relative price changes. Along this line, two much used forms are the so-called translog cost share specification and multi-nominal logit specification. The former is derived from static competitive cost

---

[2] See, for example, Berndt and Wood (1975) and Chambers (1988) for the derivation of this parameter.

[3] See, for example, Chambers (1988) and Varian (1992) for a further discussion of the properties of these function forms.

minimization under a translog unit cost function.[4] According to the translog cost share specification, fuel choice depends on own-prices and cross-prices of competing fuels. The degree to which fuels may be substituted for one another is measured by the Allen partial elasticities of substitution (AES). The AES are not constrained to be constant but may vary with the values of actual fuel cost shares. The higher the fuel cost shares, the lower the cross-price elasticities are and vice versa.

Unlike the translog cost share specification, the multi-nominal logit specification is not based on assumptions of cost-minimization, but on a probabilistic choice model instead. It assumes that the own-prices and cross-prices of competing fuels and other institutional and technological variables affect the probability of fuel choice. In practice, this implies that this logit specification can take the form of either relating the ratio of the share of two or more competing fuels to their prices or including other variables that affect the probability of fuel choice.[5]

It should be noted that both translog cost share specification and multi-nominal logit specification are estimated for thermal power generation, and do not allow substitution between fossil fuels and non-fossil ones, thus limiting the investigation of the possibilities of switching from high-carbon fossil fuels and technologies towards carbon-free counterparts for coping with the carbon emissions limits. It is of policy relevance to do such an investigation, although it is unlikely that a massive non-fossil fuel expansion could take place in response to price changes over the next 15 to 20 years (Hoeller and Coppel, 1992).

8.2.2 Optimization approach

The optimization model explicitly lists alternative electricity supply technologies, with each being characterized by its initial investment, transmission and distribution cost, fixed and variable operating and maintenance costs, generation efficiency, plant utilization, plant lifetime, construction lead times, fuel requirements, dates of introduction, maximum rates of expansion and emissions coefficients. The optimization model is usually of an intertemporal structure and thus allows for interactions between periods for the introduction of newly-emerging power supply technologies. Once each alternative in a 'shopping list' is characterized, the optimization model simulates the competition among fuels and available supply technologies, and chooses the most cost-efficient mix of technologies and fuels to meet the exogenously-determined electricity demand if the minimization of discounted cost over the entire planning horizon is chosen as the objective function.

---

[4] See, for example, Berndt and Wood (1975) and Smith (1989) for a derivation of the translog cost share specification.

[5] See, for example, Edmonds and Reilly (1983) and Walker and Birol (1992) for the applications of multi-nominal logit specification.

8.3   The MARKAL model: theory and modifications

For the cost-effective analysis of carbon abatement options in China's power sector, I have developed a technology-oriented optimization model for power system expansion planning. This model has been adapted from the MARKAL model (Fishbone et al., 1983). It chooses the minimization of discounted cost over the entire planning horizon as its objective function and incorporates a number of power-related constraints adopted by MARKAL. The model is programmed in GAMS, a widely available (non)linear programming software package.

Although developing the power planning model has been motivated by the fact that using MARKAL as a tool requires costly additional facilities,[6] the equations in MARKAL are of a general nature. That is why I will give a comprehensive description of MARKAL. The purpose is not only to contribute to a better understanding of the whole energy system on which MARKAL is based,[7] but also to shed light on those power-related elements that form our power system expansion planning model. The description is based on related literature, including Fishbone et al. (1983), Stocks and Musgrove (1984), Kleemann and Wilde (1990), Kypreos (1990), Berger et al. (1991), Goldstein (1991), Zhang (1992).

MARKAL, an acronym for MARKet ALlocation, is an optimization, dynamic, technology-oriented and demand-driven large scale linear programming model based on representing a physical flow of the energy supply system. Initially it was developed to explore technology options and costs for meeting demands for energy by Brookhaven National Laboratory (BNL) in the United States and Kernforschungsanlage Jülich (KFA) in Germany, within the framework of the International Energy Agency. MARKAL has been expanded to accommodate

---

[6]   At the time of undertaking the research, MARKAL was written in OMNI, a character-oriented language. It was implemented using OMNI matrix generation and reporting software (from Haverly Systems), XPRESS-MP optimizer (from Dash Associates) and XPRESS-OMNI Interface (from Haverly Systems). This limited to some extent access to it. Because of this, efforts had been made to rewrite MARKAL in GAMS by the Energy Technology Systems Analysis Programme of the International Energy Agency (IEA-ETSAP). Currently, by incorporating an explicit representation of the uncertainties through stochastic programming techniques, MARKAL is even used to address uncertainties surrounding greenhouse gas emissions from energy systems. See various issues of the IEA-ETSAP Newsletter and Fragnière and Haurie (1995).

[7]   Current literature on the description of MARKAL is either unreadable or too condensed, a view shared by Kypreos (1990) who states that the complete description of MARKAL given in the User's Guide (Fishbone et al., 1983) is very difficult to access.

issues surrounding emissions of pollutants, and is used as a tool for the IEA-ETSAP.

Within MARKAL, the energy system is divided into four technology classes: energy sources, process technologies, conversion technologies and demand devices. These classes are represented explicitly in the model in terms of their technical, economic and environmental characteristics. On the supply side, primary energy carriers are first extracted from resources, then transformed by supply technologies into a variety of secondary energy carriers such as fuels, electricity and low-temperature heat (also called district heat). MARKAL allows primary and/or secondary energy carriers to be imported or exported. Finally, all these energy carriers available to the system by domestic production and imports are consumed by a number of end-use demand devices that compete to satisfy various socio-economic needs usually expressed as useful or final energy demands in each sector of the economic system. A set of technologies related to extracting, converting, transporting and consuming energy constitute the reference energy system (RES) of the country in question. The RES also specifies the various energy carriers that are linked with these technologies.

On representation of these technologies and energy carriers in a 'shopping list', MARKAL simulates the competition among fuels and available technologies, and produces, from the perspective of the whole system, the cost-efficient energy system subject to a set of constraints such as energy carrier balances, demand balances, allowable emissions, resources availability, technology capacity and plant maintenance scheduling if the minimization of discounted cost over the entire planning horizon is chosen as the objective function.[8]

8.3.1 Technology description

MARKAL is a technology-oriented model. As shown in Figure 8.1,[9] the RES on which MARKAL is based includes a number of energy supply and end-use technologies available during the period under consideration, which are divided into four broad categories (classes) according to the energy carriers upon which they operate:

a) Energy sources - also called resource (production) technologies. The sources of supply cover all means by which energy can enter or leave the system. This includes mining, imports, exports, renewables and stockpiling of energy carriers, the last one allowing energy carriers produced in one period to be consumed in a subsequent period.

---

[8] There are also other choices of the objective functions in MARKAL. See Section 8.3.4 for a further discussion.

[9] This is only for illustration. In reality, the RES in question is much more complicated than this one.

b) Conversion technologies - all types of load-dependent plants generating electricity or low-temperature heat or both. Clearly, a class of all coupled electricity and low-temperature heat production plants is a subset of the class. MARKAL models two types of coupled production plants: pass-out turbine and back-pressure turbine. The first type generates heat by passing steam out of the low stage turbine to a heat exchanger and thus to the district heating grid. It can operate following the load demand for heat by reducing part of the electricity production. The second type generates electricity and heat at a constant ratio and cannot do otherwise.

c) Process technologies - all load-independent processes converting one energy carrier into another; and

d) Demand devices - all devices consuming energy carriers to meet the end-use demands such as motive power, space heat, transportation, and so on.

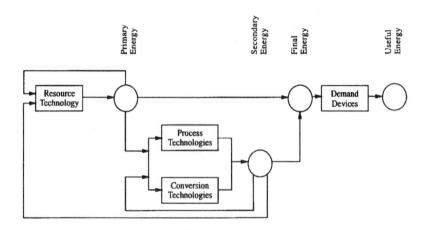

Source: Adapted from Kleemann and Wilde (1990).

Figure 8.1   *An illustrative reference energy system*

The description of each technology requires technological, economic and environmental parameters. For each resource and resource technology, the parameters required include resource cost, period of first availability, environmental emission coefficients, and security weight. For each process, conversion technology, and demand device, the requisite parameters include the following: technical lifetime, period of first availability, availability factor, energy efficiency, costs of investment and fixed and variable operation and maintenance, and environmental emission coefficients. These data are collected, assembled and cast into a form susceptible of being included in the MARKAL database through the MARKAL Users Support System (MUSS), which is a relational database management system designed specifically to facilitate the use of the MARKAL model (Goldstein, 1991).

8.3.2 Endogenous variables

The MARKAL model contains four types of endogenous variables representing investment, capacity and production of a given technology, and activity of an energy source. These variables characterize the various forms of energy carriers from the sources of supply to the end-use devices, the sources of energy, and the capacities of energy technologies modelled in MARKAL and their levels of activity.

8.3.2.1 Source activity variables
Energy carriers, excluding electricity and low-temperature heat, are mined, imported, exported, stockpiled or renewable, depending on the source specified. The source variables are defined for modelling the activity (production) of each energy-carrier resource. In Section 8.3.3, the variables represented by $ACT_{skp}(t)$ define the annual activity of energy carrier k from source s and location p in period t.

Electricity can enter or leave the system by imports or exports. In Section 8.3.3, two variables, $IMPELC_p(t)$ and $EXPELC_p(t)$, are defined to represent the annual import and export of electricity from location p in period t respectively.

Unlike electricity, imports and exports of low-temperature heat are not allowed and hence no source variables are needed.

8.3.2.2 Capacity and investment variables
Each technology is described by the capacity and investment variables for each period, with investments being equal to newly-built capacity in that period. In Section 8.3.3, these two variables are represented by $C_i(t)$ and $I_i(t)$ respectively, the first defining the capacity of technology i installed available annually in period t, and the last one defining the newly-installed capacity of technology i at beginning of period t.

8.3.2.3 Production variables
Demand devices do not have production variables because their activity is assumed to be strictly proportional to the capacity, thus requiring only investment variables and capacity variables to model such technologies. In contrast, process and conversion technologies are modelled in MARKAL by production variables together with investment variables and capacity variables. The actual production is bounded by capacity and an annual availability factor.

a) Annual production variable for each process technology which is defined by $ACT_i(t)$ in Section 8.3.3. It represents the annual production of process i in period t.

b) Variable for seasonal production of low-temperature heat by each heating plant. The variable represented by $LTH_{iz}(t)$ in Section 8.3.3 describes the annual low-temperature heat production by heating plant i in period t for season z.

c) Variable for seasonal and diurnal electricity production by each electric power plant. The variable represented by $ELC_{izy}(t)$ in Section 8.3.3 describes the annual electricity production by technology i in period t for season z and time of day y. For coupled production pass-out turbines, variables for electricity production are defined for situations in which electricity and low-temperature heat are not produced proportionally. In addition, a seasonal and diurnal heat production variable is defined; the variables represented by $LTH_{izy}(t)$ in Section 8.3.3 describe the annual low-temperature heat production by coupled-production pass-out turbine i in period t for season z and time of day y.

d) For conversion technologies with explicitly user-defined scheduled maintenance requirements, seasonal maintenance variables are defined for allocating the unavailability of the capacity installed due to a scheduled outage. In Section 8.3.3, the variables represented by $M_{iz}(t)$ describe the scheduled maintenance for conversion technology i in period t for season z.

8.3.3 Constraints

A list of constraints incorporated in MARKAL is given below:

☐   Fossil, nuclear and renewable fuel balances
☐   Electricity balance seasonal and diurnal
☐   Electric base load constraints
☐   Electric peak load constraints
☐   Seasonal low-temperature heat balances
☐   District heat peak load constraints
☐   Capacity interperiod transfer constraints
☐   Technology capacity growth constraints
☐   Cumulative energy resource availability
☐   Energy-carrier resource growth constraints
☐   Demand balances
☐   Electricity production capacity constraints
☐   District-heating production capacity constraints
☐   Process-utilization capacity constraints
☐   Scheduled maintenance for conversion technologies
☐   Output-variable process balances
☐   Total investment growth constraints
☐   Bounds on production of conversion technology
☐   Emission constraints
☐   Ratio constraints

In the following subsections, I will describe in some detail the main categories of these constraints in MARKAL, which tie the variables together. The endogenous variables and classes are written in capital letters, while the coefficients and bounds (also called the exogenous variables since they are exogenously determined) are written in lower-case letters. The coefficients characterizing a specific technology may change in each period to reflect ongoing technology development. The meaning of a symbol is explained on first use. With respect to symbolical order, classes are first defined, followed by indices, endogenous variables and parameters. Moreover, in doing so, I follow the notions of variables and coefficients used in the MARKAL *User's Guide* (Fishbone *et al.*, 1983) as much as possible in order to help readers read the original MARKAL *User's Guide*.

8.3.3.1 Flow balance constraints

Three sets of energy carrier balances ensure that for each energy carrier the sum of availability to the energy system is no less than the total amount consumed by the system. The first set of energy carrier balances applies to the annual production and consumption of fossil and nuclear fuels, the second set to the seasonal and diurnal electricity production and consumption, and the third one to seasonal production and consumption of low-temperature heat.

a) Fossil and nuclear fuel balances

$$\sum_{p \in P} MIN_{kp}(t) \times te_k + \sum_{p \in P} IMP_{kp}(t)$$

$$\geq \sum_{p \in P} EXP_{kp}(t) + \sum_{i \in PRC} ACT_i(t) \times inp_{ki}(t)$$

$$+ \sum_{i \in DMD} C_i(t) \times capunit_i \times cf_i(t) \times ma_{ki}(t) \ / \ eff_i(t)$$

$$+ \sum_{i \in ELA} \sum_{z \in Z} \sum_{y \in Y} ELC_{izy}(t) \times inp_{ki}(t)$$

$$+ \sum_{i \in HPL} \sum_{z \in Z} LTH_{iz}(t) \times inp_{ki}(t)$$

$$+ \sum_{i \in POT} \sum_{z \in Z} \sum_{y \in Y} LTH_{izy}(t) \times ceh_{izy} \times inp_{ki}(t) \ / \ elm_i \qquad \forall k \in ENC, \ \forall t \in T$$

The left-hand side of fuel balances represents the sum of availability of a given energy carrier by mining and imports, while the right-hand side represents the sum of the exports of the energy carrier and the amount of the energy carrier consumed by all process technologies, all demand devices, all electric conversion technologies, all heating plants, and by all coupled-production pass-out turbines.

where
BPT = Class of all coupled-production plants with back-pressure turbines.
POT = Class of all coupled-production plants with pass-out turbines.

CPD	=	Class of all conversion technologies generating both electricity and low-temperature heat. It is a union of the two sets *POT* and *BPT*.
ELA	=	Class of all electricity generating technologies.
HPL	=	Class of all technologies generating only low-temperature heat.
CON	=	Class of all plants generating electricity and/or low-temperature heat. It is a union of the two sets *ELA* and *HPL*.
DMD	=	Class of all end-use demand technologies.
ENC	=	Class of all energy carriers, excluding electricity and low-temperature heat.
PRC	=	Class of all process technologies.
TCH	=	Class of all technologies. It is a union of the three sets *CON*, *PRC* and *DMD*.
P	=	Class of all locations of an energy carrier.
T	=	Class of all periods of the entire planning horizon. It is also defined as the number of periods in the entire planning horizon.
Y	=	Class of daily divisions (*d* refers to day division and *n* to night division).
Z	=	Class of seasonal divisions (*w* refers to winter, *u* to summer and *m* to intermediate).
i	=	Index for technologies.
k	=	Index for energy carriers.
p	=	Index for locations or destinations of an energy carrier.
t	=	Index for time periods.
y	=	Index for daily divisions.
z	=	Index for seasonal divisions.
$ACT_i(t)$	=	Annual production (activity) of process *i* in period *t*.
$C_i(t)$	=	Capacity of technology *i* installed available annually in period *t*.
$ELC_{izy}(t)$	=	Annual electricity production by plant *i* in period *t* for season *z* and time of day *y*.
$EXP_{kp}(t)$	=	Annual export of energy carrier *k* to destination *p* in period *t*.
$IMP_{kp}(t)$	=	Annual import of energy carrier *k* from location *p* in period *t*.
$LTH_{iz}(t)$	=	Annual low-temperature heat production by heating plant *i* in period *t* for season *z*.
$LTH_{izy}(t)$	=	Annual low-temperature heat production by coupled-production pass-out turbine *i* in period *t* for season *z* and time of day *y*.
$MIN_{kp}(t)$	=	Annual mining of energy carrier *k* extracted domestically from location *p* in period *t*.

$capunit_i$	=	Factor converting units of capacity of technology i to units of its annual production (e.g. 31.536 PJ/GW).
ceh_{izy}	=	Ratio of electricity lost to low-temperature heat generated for coupled-production pass-out turbine i during season z and time of day y.
$cf_i(t)$	=	Average utilization factor of capacity of demand device i installed in period t.
$eff_i(t)$	=	Efficiency of demand device i in period t.
elm_i	=	Allowable electric loss for coupled-production pass-out turbine i.
$inp_{ki}(t)$	=	Input of energy carrier k per unit of production of technology i in period t.
$ma_{ki}(t)$	=	Market allocation of energy carrier k to demand device i in period t.
te_k	=	Transmission and distribution efficiency for energy carrier k.

b) Electricity balances In MARKAL, the year is divided into six time divisions which comprise three seasons - winter, summer and intermediate (spring and autumn) - and two daily divisions - day and night. Since electricity balances are daily-based in the model, six balance constraints are required in each period.

$$\sum_{p \in P} IMPELC_p(t) \times qhr_{zy} \times te + \sum_{i \in ELA} ELC_{izy}(t) \times te$$

$$+ \sum_{i \in POT} LTH_i(t) \times te \times ceh_{izy} \times (1 - elm_i) / elm_i$$

$$\geq \sum_{p \in P} EXPELC_p(t) \times qhr_{zy}$$

$$+ \sum_{i \in PRC} ACT_i(t) \times inpelc_i(t) \times qhr_{zy}$$

$$+ \sum_{i \in DMD} C_i(t) \times capunit_i \times maelc_i(t) \times \sum_{j \in DM} fr_{jizy} \times cf_i(t) / eff_i(t)$$

$$+ \sum_{s \in SRC} \sum_{k \in ENC} \sum_{p \in P} ACT_{skp}(t) \times inpelc_{skp}(t) \times qhr_{zy}$$

$$+ \sum_{i \in STG} ELC_{izd}(t) \times inpelcs_i(t) \times te \qquad \forall z \in Z, \ \forall y \in Y, \ \forall t \in T$$

The left-hand side of electricity balances represents the sum of the imports of electricity and the amount of electricity generated by all electric conversion technologies plus heat-production-dependent electricity production by all coupled-production pass-out turbines, while the right-hand side represents the sum of the exports of electricity and the amount of electricity consumed by all process technologies, all demand devices, energy-carrier resource activities, and by all storage conversion technologies.

where

DM	=	Class of all end-use energy demand categories.
SRC	=	Class of all sources of supply of energy carriers. It includes mining (MIN), imports (IMP), exports (EXP), renewables (RNW) and stockpiling (STK) of energy carriers.
STG	=	Class of all electricity storage technologies, for example, hydro-electric pumped storage.
j	=	Index for energy demand categories.
s	=	Index for sources of an energy carrier.
$ACT_{skp}(t)$	=	Annual activity of energy carrier k from source s and location p in period t.
$EXPELC_p(t)$	=	Annual export of electricity from location p in period t.
$IMPELC_p(t)$	=	Annual import of electricity from location p in period t.
fr_{jizy}	=	Fraction of demand category j to which demand device i belongs that occurs in season z and time of day y.
$inpelc_i(t)$	=	Electricity input per unit of production of process i in period t.
$inpelc_{skp}(t)$	=	Electricity input required to make available to the system primary energy carrier k from source s and location p in period t.
$inpelcs_i(t)$	=	Electricity input during the night required to generate one unit of electricity during the day for storage technology i in period t.
$maelc_i(t)$	=	Market allocation of electricity to demand device i in period t.
qhr_{zy}	=	Fraction of year occurring in season z and time of day y.
te	=	Transmission and distribution efficiency of the electricity grid.

c) Low-temperature heat balances Unlike electricity, low-temperature heat balances in MARKAL are on a seasonal basis. Therefore, only three balance constraints are needed in each time period. Moreover, low-temperature heat is only used by demand devices and imports and exports of it are not allowed.

$$\sum_{i\in HPL} LTH_{iz}(t) + \sum_{i\in POT}\sum_{y\in Y} LTH_{izy}(t)\times th_{izy} + \sum_{i\in BPT}\sum_{y\in Y} ELC_{izy}(t)\times th_{izy}/reh_i$$

$$\geq \sum_{i\in DMD}\left[C_i(t)\times capunit_i \times malth_i(t)\times\sum_{y\in Y}\sum_{j\in DM} fr_{jizy}\times cf_i(t)/(dh_{iz}\times eff_i(t))\right]$$

$$\forall z \in Z, \ \ \forall t \in T$$

The left-hand side of heat balances represents the sum of production of low-temperature heat from all heating plants, all coupled-production pass-out turbines, and from all coupled-production back-pressure turbines, while the right-hand side defines the consumption of low-temperature heat by all demand devices.

where

dh_{iz}	=	Distribution efficiency of low-temperature heat to demand device i in season z.
$malth_i(t)$	=	Market allocation of low-temperature heat to demand device i in period t.
reh_i	=	Ratio of electricity produced to heat produced for coupled-production back-pressure turbine i.
th_{izy}	=	Transmission efficiency of low-temperature heat from coupled-production plant i in season z and time of day y.

8.3.3.2 Energy-carrier resource constraints

Two types of upper bounds are imposed on the activity of energy carriers which are mined (*MIN*), imported (*IMP*), exported (*EXP*), stockpiled (*STK*) or renewable (*RNW*). The class *SRC* specifies the way in which these energy carriers are made available to the system.

a) Energy-carrier resource growth constraints The resource growth constraints, if utilized, limit the growth in activity of an energy-carrier resource between periods, and allow for an annual growth of activity according to the activity in the previous period plus an initial user-specified value since the constraints would otherwise not allow growth from zero.

$$ACT_{skp}(t) \leq ACT_{skp}(t-1) \times ag_{skp}(t-1)^{ny} + ga_{skp}$$

$$\forall s \in SRC, \ \forall k \in ENC, \ \forall p \in P, \ \forall t \in T$$

where

$ag_{skp}(t-1)$	=	Average allowable annual growth factor in the activity of energy carrier k from source s and location p during period $(t-1)$.
ga_{skp}	=	Activity increase allowed for energy carrier k from source s and location p in addition to specified growth rate.
ny	=	Number of years in each period.

b) Cumulative energy-carrier resource bound Compared with the above-mentioned resource growth constraints that hold individually in each period, another possible constraint on resources is that total availability over all periods of the entire time horizon is less than or equal to a user-defined cumulative value. In the case of mining, both extraction and *in-situ* consumption of energy carriers are included.

$$\sum_{t \in T} ACT_{skp}(t) \times ny$$

$$+ \sum_{t \in T} \sum_{i \in ELA} \sum_{z \in Z} \sum_{y \in Y} ELC_{izy}(t) \times ny \times rf_{ip}(t)$$

$$+ \sum_{t \in T} \sum_{i \in HPL} \sum_{z \in Z} LTH_{iz}(t) \times ny \times rf_{ip}(t)$$

$$+ \sum_{t \in T} \sum_{i \in POT} \sum_{z \in Z} \sum_{y \in Y} LTH_{izy}(t) \times ny \times rf_{ip}(t) \times ceh_{izy} \,/\, elm_i$$

$$+ \sum_{t \in T} \sum_{i \in PRC} ACT_i(t) \times ny \times rf_{ip}(t) \leq cum_{skp}$$

$$\forall s \in SRC, \quad \forall k \in ENC, \quad \forall p \in P$$

The left-hand side of cumulative resource bound represents, over all periods of the entire planning horizon, the sum of total activity of the indexed energy-carrier resource and the amount of *in-situ* consumption of a given, indexed energy carrier by all electric conversion technologies, all heating plants, all coupled-production pass-out turbines, and by all process technologies.

where

| $rf_{ip}(t)$ | = | *in-situ* consumption of fossil energy carrier from location p per unit of production of technology i in period t. |
| cum_{skp} | = | Total availability of energy carrier k from source s and location p over all periods of the entire time horizon. |

8.3.3.3 Capacity constraints
The capacity constraints determine the necessary capacities installed and the newly-built capacities in each period.

a) Capacity interperiod transfer constraints The dynamic character of the energy system is obtained through the capacity transfer relations, which link the current capacity of a technology in period t with gradually retiring capacity installed before start of optimization and still available for use in that period within its lifetime. These relations are a set of equalities for process and conversion technologies and inequalities for demand devices. The distinction is made because demand devices have only capacity variables, while the non-demand technologies have both capacity and activity variables.

$$C_i(t) \leq \sum_{\tau = t - l_i + 1}^{t} I_i(\tau) + resid_i(t) \qquad \forall i \in TCH, \quad \forall t \in T$$

where

$I_i(\tau)$	=	New capacity of technology i installed at beginning of period τ.
l_i	=	Lifetime of technology i.
$resid_i(t)$	=	Capacity installed before start of optimization available for use in period t and for which no investment cost is charged.

b) Capacity growth constraints The capacity growth constraints, if utilized, limit the growth in capacity of a technology between periods, and allow for an annual growth of capacity according to the capacity in the previous time period plus an initial user-specified value since the constraints would otherwise not allow growth from zero.

$$C_i(t) \leq C_i(t-1) \times cg_i(t-1)^{ny} + gc_i \qquad \forall i \in TCH, \forall t \in T$$

where

$cg_i(t-1)$	=	Average allowable annual growth factor in technology i during period $(t-1)$.
gc_i	=	Capacity increase allowed for technology i in addition to specified growth rate.

An alternative to restrict the rate of growth of technologies in MARKAL is by means of imposing explicit bounds on technology capacity. Upper bounds are placed on the rate of market penetration for new energy technologies that are supposed to become available at specified points in the future, and there are lower bounds to ensure that older technology will not be phased out too rapidly. Bounds on technology capacity can be written as follows:

$$C_i(t) \leq ubc_i(t) \qquad \forall i \in TCH, \quad t \geq t_i$$

$$C_i(t) \geq lbc_i(t) \qquad \forall i \in TCH, \quad \forall t \in T$$

where
$lbc_i(t)$ =   Lower bound on capacity of technology i in period t.
$ubc_i(t)$ =   Upper bound on capacity of new technology i in period t.

8.3.3.4 Demand balances
The demand constraints are generated for each energy demand category and for each period. These relations ensure that the sum of end-use output from all demand devices that contribute to a particular demand category is greater than or equal to the end-use demand for that energy carrier that is exogenously determined. This implies that MARKAL is a demand-driven model.

$$\sum_{i \in DMD} C_i(t) \times capunit_i \times out_{ji}(t) \times cf_i(t) \geq demand_j(t) \qquad \forall j \in DM, \quad \forall t \in T$$

where
$demand_j(t)$ =   End-use demand of energy category j in period t.
$out_{ji}(t)$ =   Fraction of capacity of demand device i meeting end-use demand of energy category j in period t.

8.3.3.5 Production constraints
Since process and conversion technologies have got both capacity and activity variables, these production relations ensure that the production (activity) of each technology in each period is less than or equal to its available capacity. Three types of production relations are distinguished.

a) Process-utilization capacity constraints

$$ACT_i(t) \leq C_i(t) \times capunit_i \times af_i(t) \qquad \forall t \in T, \quad \forall i \in PRC$$

b) Electricity production capacity constraints For coupled-production plants with pass-out turbines

$$ELC_{izy}(t) + LTH_{izy}(t) \times ceh_{izy} \mathbin{/} elm_i + M_{iz}(t) \times \left(\frac{qhr_{zy}}{qhr_{zd} + qhr_{zn}} \right)$$

$$\leq\ C_i(t) \times capunit_i \times qhr_{zy} \times (1 - (1 - af_i(t)) \times fo_i)$$

$$\forall z \in Z, \quad \forall y \in Y, \quad \forall t \in T, \quad \forall i \in POT$$

For electricity production plants other than those with pass-out turbines

$$ELC_{izy}(t) + M_{iz}(t) \times \left(\frac{qhr_{zy}}{qhr_{zd} + qhr_{zn}} \right)$$

$$\leq\ C_i(t) \times capunit_i \times qhr_{zy} \times (1 - (1 - af_i(t)) \times fo_i)$$

$$\forall z \in Z, \quad \forall y \in Y, \quad \forall t \in T, \quad \forall i \in BPT$$

c) District-heating production capacity constraints

$$LTH_{iz}(t) + M_{iz}(t)$$

$$\leq\ C_i(t) \times capunit_i \times \sum_{y \in Y} qhr_{zy} \times (1 - (1 - af_i(t)) \times fo_i)$$

$$\forall z \in Z, \forall t \in T, \forall i \in HPL$$

where, in constraints *a*, *b* and *c*

$M_{iz}(t)$	$=$	Scheduled maintenance for conversion technology i in period t for season z.
$af_i(t)$	$=$	Annual availability of technology i in period t.
fo_i	$=$	Fraction of annual unavailability which is defined to be forced outage.
qhr_{zd}	$=$	Same as qhr_{zy} for $y = d$.
qhr_{zn}	$=$	Same as qhr_{zy} for $y = n$.

8.3.3.6 Scheduled maintenance for conversion technologies

The utilization relation is generated for each conversion technology producing electricity and/or low-temperature heat. It allocates scheduled plant maintenance by season:

$$\sum_{z \in Z} M_{iz}(t) \geq C_i(t) \times capunit_i \times (1 - af_i(t)) \times (1 - fo_i) \quad \forall t \in T, \quad \forall i \in CON$$

8.3.3.7 Electric and district heat peak load constraints

The peak constraints ensure that there is extra capacity to meet peak requirements for electricity and low-temperature heat during the time of maximum load. As for electricity, it is assumed that this peak demand occurs during daytime in winter or summer, while as for low-temperature heat, this constraint is valid only for winter. These peak constraints differ from the electricity and district heat balances in their focus on generating capacity as opposed to electricity and low-temperature heat production.

a) Electric peak load constraints

$$\frac{te}{1+er}\left(\sum_{p\in P} IMPELC_p(t)\times fre_{pzd}\ /qhr_{zd} + \sum_{i\in ELA} C_i(t)\times capunit_i\times pk_i(t)\right)$$

$$\geq \quad \sum_{p\in P} EXPELC_p(t)\times pke_p(t)\times fre_{pzd}\ /qhr_{zd}$$

$$+\quad \sum_{i\in DMD} C_i(t)\times capunit_i\times maelc_i(t)\times \sum_{j\in DM} fr_{jizd}\times cf_i(t)\times elf_i(t)/(qhr_{zd}\times eff_i(t))$$

$$+\quad \sum_{i\in PRC} ACT_i(t)\times inpelc_i(t)\times epk_i(t)$$

$$+\quad \sum_{s\in SRC}\sum_{k\in ENC}\sum_{p\in P} ACT_{skp}(t)\times inpelc_{skp}(t)\times epk_{skp}(t) \qquad \forall t\in T,\ z=w,u$$

The left-hand side of electric peak load constraints represents, during the daytime of maximum load, the sum of the imports of electricity and the amount of electricity generated by all electric conversion technologies, while the right-hand side represents, during the corresponding time, the sum of the exports of electricity and the amount of electricity consumed by all demand devices, all process technologies, and by energy-carrier resource activities.

where

$elf_i(t)$	=	Ratio of electricity consumption at the time of peak demand to the time division average for demand device i in period t.
$epk_i(t)$	=	Fraction of electricity consumption by process i which is to be included in the peak relation in period t.
$epk_{skp}(t)$	=	Fraction of electricity consumption required to make available to the system primary energy carrier k from

source s and location p which is to be included in the peak relation in period t.

er	$=$	Capacity reserve factor for the electricity sector expressed as a fraction of the capacity required to meet electricity demand during the time division of greatest load.
fr_{jizd}	$=$	Same as fr_{jizy} for $y = d$.
fre_{pzd}	$=$	Fraction of annual electricity imports or exports from location p occurring in season z and during daytime d.
$pk_i(t)$	$=$	Fraction of capacity of generating plant i installed available to meet peak load in period t.
$pke_s(t)$	$=$	Fraction of exports of electricity from location s which is to be included in the peak relation in period t.

b) District heat peak load constraints

$$\sum_{i \in HPL} C_i(t) \times capunit_i \times pk_i(t)/(1+hr)$$

$$+ \sum_{i \in POT} C_i(t) \times capunit_i \times th_{iwd} \times elm_i \times pk_i(t)/((1+hr) \times ceh_{iwd})$$

$$+ \sum_{i \in BPT} C_i(t) \times capunit_i \times th_{iwd} \times pk_i(t)/((1+hr) \times reh_i)$$

$$\geq \sum_{i \in DMD} C_i(t) \times capunit_i \times malth_i(t) \times \frac{\sum_{j \in DM} (fr_{jiwd} + fr_{jiwn})}{qhr_{wd} + qhr_{wn}} \times cf_i(t)/(dh_{iw} \times eff_i(t)) \quad \forall t \in T$$

The left-hand side of district heat peak load constraints represents, during the daytime of maximum load in winter, the sum of production of low-temperature heat by all heating plants, all coupled-production pass-out turbines, and all coupled-production back-pressure turbines, while the right-hand side defines peak demand for low-temperature heat by all demand devices.

where

ceh_{iwd}	$=$	Same as ceh_{izy} for $z = w$ and $y = d$.
dh_{iw}	$=$	Same as dh_{iz} for $z = w$.
fr_{jiwd}	$=$	Same as fr_{jizy} for $z = w$ and $y = d$.
fr_{jiwn}	$=$	Same as fr_{jizy} for $z = w$ and $y = n$.
hr	$=$	Capacity reserve factor for the low-temperature heat sector expressed as a fraction of the capacity required to meet heat demand during the time division of greatest load.

qhr_{wd} = Same as qhr_{zy} for $z = w$ and $y = d$.
qhr_{wn} = Same as qhr_{zy} for $z = w$ and $y = n$.
th_{iwd} = Same as th_{izy} for $z = w$ and $y = d$.

8.3.3.8 Electric base load constraints

Certain electric power plants, particularly large coal-fired and nuclear power plants, operate at the same level of production during peak and off-peak time of the day, that is, are base-loaded, either for technical or economic reasons. The base load relation generated for each season ensures nightly electricity production by all base-loaded electricity generating technologies plus net nightly electricity imports not to exceed a fraction of the sum of nightly electricity production by all electricity generating technologies plus net nightly electricity imports.

$$\sum_{p \in P} IMPELC_p(t) \times (1-bl) \times qhr_{zn} \times te + \sum_{p \in P} EXPELC_p \times (bl-1) \times qhr_{zn}$$

$$+ \sum_{i \in BAS} ELC_{izd}(t) \times (1-bl) \times te \times qhr_{zn} / (qhr_{zd} + qhr_{zn})$$

$$+ \sum_{i \in BAS} LTH_{izd} \times ceh_{izd} \times ((1-elm_i)/elm_i) \times (1-bl) \times qhr_{zn} \times te/(qhr_{zd} + qhr_{zn})$$

$$- \sum_{i \in BAS} ELC_{izn}(t) \times bl \times te$$

$$- \sum_{i \in BAS} LTH_{izn}(t) \times bl \times ceh_{izn} \times te \times (1-elm_i)/elm_i \leq 0 \quad \forall z \in Z, \quad \forall t \in T$$

The first two terms on the left-hand side of electric base load constraints are related to the night fraction of electricity imports and to the night fraction of electricity exports respectively, the third term to the base load electricity at night, the fourth term to the base load electricity dependent on the production of low-temperature heat by coupled-production pass-out turbines, the fifth term to the non-base-load electricity at night, and the last term to the non-base-load electricity dependent on the production of low-temperature heat by coupled-production pass-out turbines.

where
BAS = Class of electricity generating plants which are base-loaded, that is, which operate day and night at the same level of production.
bl = Maximal fraction of total nightly electricity production that can be met by base load electricity generating technologies.

8.3.3.9 Emission constraints

For a given pollutant v, total emissions are related to the activities of energy-carrier resources and technologies and to investments in or capacities of technologies. Thus the sum of all annual emissions of pollutant v associated with the energy system in period t, $ENV_v(t)$, is given by

$$ENV_v(t) = \sum_{s \in SRC} \sum_{k \in ENC} \sum_{p \in P} ACT_{skp}(t) \times es_{skpv}(t)$$

$$+ \sum_{i \in ELA} \sum_{z \in Z} \sum_{y \in Y} ELC_{izy}(t) \times ea_{iv}(t)$$

$$+ \sum_{i \in HPL} \sum_{z \in Z} LTH_{iz}(t) \times ea_{iv}(t)$$

$$+ \sum_{i \in POT} \sum_{z \in Z} \sum_{y \in Y} LTH_{izy}(t) \times ea_{iv}(t) \times ceh_{izy}/elm_i$$

$$+ \sum_{i \in PRC} ACT_i(t) \times ea_{iv}(t) + \sum_{i \in DMD} C_i(t) \times cf_i(t) \times ec_{iv}(t)$$

$$+ \sum_{i \notin DMD} C_i(t) \times ec_{iv}(t) + \sum_{i \in TCH} I_i(t) \times ei_{iv}(t)/ny \qquad \forall v \in V, \ \forall t \in T$$

The right-hand side of emission accounting equations represents the sum of all emissions of a given pollutant resulting from energy-carrier resource activities, electricity production of electric conversion technologies, from low-temperature heat production of heating plants, production of coupled-production pass-out turbines, activities of process technologies, final energy input to demand devices, capacities installed of process and conversion technologies, and from investments in technologies.

where

V	=	Class of all pollutants.
v	=	Index for pollutants.
$ENV_v(t)$	=	Annual emissions of pollutant v in period t.
$ea_{iv}(t)$	=	Emissions of pollutant v per unit of production of technology i in period t.
$ec_{iv}(t)$	=	Emissions of pollutant v per unit of capacity of non-demand technology i installed in period t.
$ei_{iv}(t)$	=	Emissions of pollutant v per incremental unit of capacity of technology i installed in period t.
$es_{skpv}(t)$	=	Emissions of pollutant v per unit of activity of energy-carrier resource k from source s and location p in period t.

In MARKAL, the national environmental policy goals are expressed as upper bounds on total allowable emissions of different pollutants. The fixed ceiling can be attained at the least abatement costs by means of emissions trading. In this regard, there are two types of emission constraints that can be applied to identify cost-effective strategies for coping with emission limits. If the ceiling is set annually for each time period, the emission constraint can be imposed on the system as follows:

$$ENV_v(t) \le ube_v(t) \qquad \forall v \in V, \ \forall t \in T$$

where
$ube_v(t)$ = Upper bound on annual emissions of pollutant v in period t.

Alternatively, the ceiling for a given pollutant can be expressed as a constant cumulative amount over the entire time horizon. In this case the emission constraint for the pollutant is given in the form:

$$\sum_{t \in T} ENV_v(t) \times ny \le ube_v \qquad \forall v \in V$$

where
ube_v = Upper bound on total emissions of pollutant v over the entire time horizon.

8.3.3.10 Bounds on investment
The bounds prevent the matrix optimizer from creating an unrealistic pattern of investments from one period to the next.

$$INVEST(t) \le INVEST(t-1) \times smooth^{ny} \qquad \forall t \in T$$

where
$INVEST(t)$ = Total investment costs for the newly-added capacity of all technologies in period t.
$smooth$ = Allowable interperiod increase of the total investment during any period.

8.3.3.11 Ratio constraints
In addition to the standard MARKAL constraints, the ratio constraints provide users with extended flexibility in defining relations which are not available in the standard model. Therefore, the relations are unique because their structure can completely be defined by the users. Through the ratio tables, users can enter any

combination of variables and coefficients to solve modelling problems which are not addressed by other relations, provided that users do not attempt to define new vectors which are not already in the matrix.

8.3.4 Objective functions

In the MARKAL model, several generalized objective functions have been programmed. This allows a range of choices for any given run. MARKAL will simultaneously determine the optimal solution to values of all variables in such a way that they satisfy all of the constraints and the objective function set up for the system. The objective functions programmed in the model are given below:

□ *PRICE*, which is the minimization of the discounted costs of the energy system over the entire planning horizon. This objective function is most commonly used in many applications.

□ *SECURITY*, which is the minimization of the sum of cumulative utilization of any energy carrier over the entire time horizon. The most realistic case of *SECURITY* is the minimization of the cumulative oil use.

□ *ENVIRONMENT*, which is the minimization of the cumulative environmental emissions over the entire time horizon.

□ *QSLOPE*, which is the weighted combination of the discounted system costs and security. It allows for examining a trade-off between *PRICE* and *SECURITY*.

Other functions available for use as objective functions include usages of fossil energy, nuclear energy, and nonrenewable energy as well as total discounted system costs with renewable technologies having zero investment costs. Treatment of the last objective is to emphasize the role of renewable sources of energy. Clearly, this is unrealistic.

In what follows, I will describe the *PRICE* function in more detail because it has been chosen as the objective function in the cost-effective analysis of carbon abatement options in China's power sector.

8.3.4.1 The *PRICE* function

The *PRICE* function represents the discounted sum of all costs of the energy system over the entire planning horizon. It can be considered as consisting of three parts: investment costs, annual costs and salvage values. The distinction is made for the first two parts because investments are assumed to take place at the beginning of each time period, while the annual costs are assumed to occur in each year of each time period. The third part handles the salvage values that provide the correction to the end-effect problem which occurs in dynamic (time-dependent) linear programming models. The correction reduces the investment cost of a technology when part of its technical lifetime will extend beyond the last time period.

a) Investment costs Investment costs include all capital investments required to make the incremental capacity operational at the beginning of the time period for all technologies plus any transmission and/or distribution investment costs of conversion technologies. No account is taken of the costs already invested in the energy system which existed at the start of the optimization period.

$$INVEST(t) = \sum_{i \in TCH} invcost_i(t) \times I_i(t) \qquad \forall t \in T$$

where

$invcost_i(t)$ $\quad = \quad$ Total investment cost per incremental unit of capacity of technology i in period t.

b) Annual costs Annual costs include costs of mining, imports, exports, renewables and stockpiling of energy carriers; fuel-delivery costs; fixed operating and maintenance (O&M) costs of all technologies; and variable O&M costs of process and conversion technologies. There are no variable O&M costs of demand technologies because they do not have activity variables. Thus, total annual system costs in period t, $ANNCOST(t)$, are represented by

$$ANNCOST(t) = \sum_{i \in TCH} C_i(t) \times fixom_i(t)$$

$$+ \sum_{i \in PRC} ACT_i(t) \times \left(\sum_{k \in ENT} del_{ki}(t) \times inp_{ki}(t) + varom_i(t) \right)$$

$$+ \sum_{i \in DMD} C_i(t) \times \left(\sum_{k \in ENT} del_{ki}(t) \times capunit_i \times ma_{ki}(t) \right) \times cf_i(t)/eff_i(t)$$

$$+ \sum_{i \in ELA} \sum_{z \in Z} \sum_{y \in Y} ELC_{izy}(t) \times \left(\sum_{k \in ENC} del_{ki}(t) \times inp_{ki}(t) + varom_i(t) \right)$$

$$+ \sum_{i \in HPL} \sum_{z \in Z} LTH_{iz}(t) \times \left(\sum_{k \in ENC} del_{ki}(t) \times inp_{ki}(t) + varom_i(t) \right)$$

$$+ \sum_{i \in CPD} \sum_{z \in Z} \sum_{y \in Y} LTH_{izy}(t) \times ceh_{izy} \times (1/elm_i - 1) \times \left(\sum_{k \in ENC} del_{ki}(t) \times inp_{ki}(t) + varom_i(t) \right)$$

$$+ \sum_{k \in ENC} \sum_{p \in P} cost_{kp}(t) \times MIN_{kp}(t)$$

$$+ \sum_{k \in ENC} \sum_{p \in P} cost_{kp}(t) \times IMP_{kp}(t) - \sum_{k \in ENC} \sum_{p \in P} price_{kp}(t) \times EXP_{kp}(t)$$

$$+ \sum_{p \in P} coste_p(t) \times IMPELC_p(t) - \sum_{p \in P} pricee_p(t) \times EXPELC_p(t) \qquad \forall t \in T$$

where

ENT	=	Class of all energy carriers. It is a union of the set *ENC*, electricity and low-temperature heat.
ANNCOST(t)	=	Total annual system costs in period t. The values are not discounted for reporting purposes.
$cost_{kp}(t)$	=	Unit cost of energy carrier k from location p in period t.
$coste_p(t)$	=	Unit cost of electricity from location p in period t.
$del_{ki}(t)$	=	Delivery cost of energy carrier k to technology i in period t.
$fixom_i(t)$	=	Annual fixed operating and maintenance costs associated with the capacity of technology i installed in period t and charged regardless of the plant utilization.
$price_{kp}(t)$	=	Unit price of energy carrier k to destination p in period t.
$pricee_p(t)$	=	Unit price of electricity to destination p in period t.
$varom_i(t)$	=	Annual variable operating and maintenance costs of the non-demand technology i in period t. These costs include items, except energy, which are proportional to the production level.

c) Salvage values Salvage values are associated with the annualized capital recovery factor crf_l. This factor represents the amount of money needed per annum to cover all fixed charges of an initial investment within a given number of years l. Given an annual discount rate, crf_l can be expressed by

$$crf_l = \frac{\alpha}{1 - (1 + \alpha)^{-l}}$$

where

α	=	Annual discount rate.
crf_l	=	Capital recovery factor of an initial investment within a given number of years l.

For newly-built capacity of technology i, its initial investments $invcost_i(t) \times I_i(t)$ will be covered over all subsequent years until the lifetime of the technology expires. When the planning horizon of analysis expires before reaching the end of the lifetime of a technology, there are extra charges for the periods beyond the entire horizon. Salvage values represent these extra charges and should be subtracted from the system costs. Discounted back to the period the investments have been made, salvage values are as follows:

$$SALVAGE(t) = \sum_{i \in TCH} invcost_i(t) \times I_i(t) \times \frac{crf_{ny \times l_i}}{crf_{ny \times (t + l_i - T - 1)}} \times (1 + \alpha)^{-ny \times (T - t + 1)}$$

where

SALVAGE(t)	=	Total salvage values for the investments made in period t.
$crf_{ny \times li}$	=	Same as crf_l for $l = ny \times l_i$.
$crf_{ny \times (t+li-T-1)}$	=	Same as crf_l for $l = ny \times (t + l_i - T - 1)$.

The *PRICE* function is the sum of the three above-mentioned functions that are discounted back to the beginning of the planning horizon. There is a difference in discounting the annual costs and other costs. The annual costs in each time period are first discounted to the beginning of the period by factor df_o, and then to the beginning of the entire horizon by factor $df(t)$. We thus have the *PRICE* function in the form:

$$\sum_{t \in T} INVEST(t) \times df(t) \; + \; \sum_{t \in T} ANNCOST(t) \times df_o \times df(t) \; - \; \sum_{t \in T} SALVAGE(t) \times df(t)$$

which is to be minimized subject to a set of constraints given in Section 8.3.3. where

$$df_o = \sum_{k=0}^{k=ny-1} (1 + \alpha)^{-k}$$

$$df(t) = (1 + \alpha)^{-ny \times (t-1)} \qquad \forall \; t \in T$$

This completes the summarized description of the standard MARKAL model. In the following section, I will describe some modifications made to MARKAL.

8.3.5 Modifications of the Standard MARKAL Model

Currently, MARKAL is being used in more than 20 countries, most of which are also active members of the Energy Technology Systems Analysis Programme (ETSAP) of the IEA.[10] According to the questions to be answered, the standard version of MARKAL has been modified by the MARKAL users along the following lines:

(a) Representation of material flows;
(b) Regionalization of MARKAL;
(c) Linkage to macroeconomic models.

In general, these modifications reflect that the MARKAL users attempt to give a comprehensive and reliable analysis of various options. In Sections 4.3 and 4.6,

---

[10] The IEA-ETSAP Newsletter, No. 4, December 1994.

linkage between MARKAL and macroeconomic models has been discussed because of its relevance to this study, whereas a discussion of representation of material flows in MARKAL is beyond the scope of this book and is therefore not touched on here.[11] In the following, I discuss briefly the rationale for regionalization of MARKAL, which is relevant to the following discussion on limitations in the representation of the electricity sector in MARKAL.

Regionalization of an energy system is considered realistic if there is a great regional diversity. Regionalization of energy demand was already possible with the standard version of MARKAL since the users are free to choose the number and definitions of demand subsectors (Berger *et al.*, 1991).[12]

On the supply side of the standard MARKAL model, it is also quite easy for most energy carriers from different regions, since the names of these energy carriers are basically chosen by the users. The problem, however, arises when modelling multiple electricity and low-temperature heat energy forms and grids. This problem arises simply because there is one single name that is built-in in MARKAL for each such energy carrier, namely, $ELC_{izy}(t)$ and $LTH_{izy}(t)$ respectively. Thus, the essence of regionalization of MARKAL is to enable multiple electricity and low-temperature heat energy forms and grids to be defined and hence to generate the equations for multiple grids.[13] Of course, regionalization of supply of electricity and district heat in MARKAL also makes possible a finer modelling of demand for these two energy carriers, given the fact that demand technologies may consume electricity and low-temperature heat generated by different grids.

Regionalization of the electricity sector in MARKAL has first been implemented in the study of the Australian electricity sector (Stocks and Musgrove, 1984). It is further extended to include the district heating sector, and

---

[11] It has been shown that when MARKAL represents material flows from its primary production to waste disposal as well as energy flows, additional cost-effective options for reducing CO_2 emissions are found. For further exploring the rationale for modification (a), see Zhang (1992).

[12] See Stocks and Musgrove (1984) for an illustrated discussion of demand regionalization.

[13] This has been exemplified by the Australian regionalized version of MARKAL, called MENSA. The essence of MENSA is not to specify new types of equations but to allow existing equations to be repeated for different grids. In MENSA, for example, a class of electricity grids, E, is defined in a study involving the southeastern Australian states, this class consists of N for New South Wales, V for Victoria, and S for South Australia. Instead of one electricity form $ELC_{izy}(t)$, one for each state is allowed: $NLC_{izy}(t)$ for New South Wales, $VLC_{izy}(t)$ for Victoria, and $SLC_{izy}(t)$ for South Australia (Stocks and Musgrove, 1984).

its implementation is exemplified by the investigation of the Canadian energy system (Berger *et al.*, 1991).

In what follows, I propose some further modifications needed to make MARKAL suitable for detailed analysis of electric utility systems.

I start with limitations in the representation of the electricity sector in MARKAL. MARKAL is originally designed to assess impacts of new technologies on the energy system, national or regional, in order to indicate broad trends in supply and demand not only for electricity but for all fuels. MARKAL analyses the electricity sector as part of the overall energy system competing with other sectors for fuel resources to meet end-use demands. This leaves MARKAL unsuitable for detailed analysis of electric utility systems as compared with the more detailed utility simulation models, such as WASP (Wien Automatic System Planning Package) and EGEAS (Electric Generation Expansion Analysis System).[14] The reason for this can be explained by the following observations.

First, power plants are usually not identified individually but are aggregated into plant types in MARKAL. The treatment is unsuitable for analysing daily operation of individual plants and system expansion plans in fine detail.

Second, perhaps more importantly, MARKAL does not adequately address technical considerations vital to utility operations. For example, no account is taken of the stability requirements for the grid. Although allowance is made for both forced and scheduled outages for power plants, this is on an average yearly basis and does not account for a combination of outages in the system during the year (Stocks and Musgrove, 1984). Also, consideration of utility operations is not fine enough to prevent unrealistic production for meeting the peak load.

Third, no account is taken of the location of loads and plants in MARKAL. This is a common failure inherent in many models of nationally-aggregated electricity systems. As a consequence, an additional unit of electricity, if needed, is made available to the system from a nationally-aggregated system rather than the plants nearest to the loads, although the latter may be the case in reality.

Given these limitations inherent in MARKAL, some modifications need to be made if MARKAL is used to deal with electricity system expansion planning. Regionalization of MARKAL as discussed above can overcome the third problem. A probabilistic treatment of forced outages is useful for meeting the stability requirements. But if MARKAL is used to address the issue of long-term electricity system expansion planning and to complement the more detailed utility simulation models, such a treatment of outages in the standard MARKAL model is considered acceptable given too many uncertainties involved in the long-term

---

[14] WASP was originally developed by the Tennessee Valley Authority and later progressively improved by the International Atomic Energy Agency, while EGEAS was developed by the US Electric Power Research Institute. For a further discussion of these two models, see, for example, Codoni *et al.* (1985).

planning. In what follows, I will thus concentrate on preventing unrealistic utility operations for meeting the peak load.

8.3.5.1 Thermal power plant operation constraints

Fuel-fired plants, if counted on for peak and reserve margin requirements, must operate at loads above the technically allowable minimum. Their annual utilization would otherwise be reduced to the level that is technically difficult to achieve if these plants were utilized to meet the peak load. In MARKAL, the minimum load, however, is not taken into account. Thus, an extra constraint is required to prevent the model from overestimating operation of these plants during the time division of greatest load.

$$\sum_{z \in Z} \sum_{y \in Y} ELC_{izy}(t) \geq C_i(t) \times mc_i \times capunit_i \times af_i(t) \qquad \forall i \in ELA, \ \forall t \in T$$

where

mc_i = Ratio of the allowable minimum load to the rated power of fuel-fired plant i.

8.3.5.2 Hydropower plant operation constraints

The production of hydropower plants of the storage type depends on the availability of both plants and reservoirs. While plant availability in each period is represented explicitly in MARKAL by $af_i(t)$, no account is taken of the reservoir availability. As a consequence, the storage type plants may operate at loads much higher than the designed specifications. To prevent the model from overestimating production, the following extra restraint on reservoir availability is needed:

$$\sum_{z \in Z} \sum_{y \in Y} ELC_{izy}(t) \leq C_i(t) \times capunit_i \times ar_i \qquad \forall i \in ELA, \ \forall t \in T$$

where

ar_i = Annual availability of reservoir associated with storage type plant i.

Moreover, how hydropower plants operate during the time of greatest load depends on the reservoir availability in winter, and should be reflected in MARKAL to prevent the model from overcounting on hydropower plants for meeting the peak load. Thus, the extra restraint on hydropower plant operation is needed and can be expressed as follows:

$$\sum_{y \in Y} ELC_{iwy}(t) \leq C_i(t) \times capunit_i \times [qhr_{wd} + qhr_{wn}] \times wa_i \quad \forall i \in ELA, \; \forall t \in T$$

where

$ELC_{iwy}(t)$ =   Same as $ELC_{izy}(t)$ for $z = w$.

wa_i =   Availability of reservoir associated with storage type plant i in winter.

8.4   Alternative low CO_2-emitting electricity supply technologies

The technologies to be considered in the power sector include:

- [ ] Coal-fired power (<200 MW)
- [ ] Coal-fired power (200 MW $\leq$ unit capacity $\leq$ 300 MW)
- [ ] Coal-fired power (>300 MW)
- [ ] Hydroelectric power (>25 MW)
- [ ] Mini-hydroelectric power (≤ 25 MW)
- [ ] Pumped storage hydroelectric power
- [ ] Nuclear power (300 MW)
- [ ] Nuclear power (600 MW $\sim$ 1000 MW)
- [ ] Imported natural gas-fired power
- [ ] Centralized solar photovoltaic (PV) power
- [ ] Decentralized solar PV power
- [ ] Wind-driven power generation
- [ ] Decentralized mini-wind power generator
- [ ] Biomass-based power generation
- [ ] Geothermal-based power generation

With respect to these alternative power supply technologies, two points have to be mentioned.

First, note that conventional coal, hydroelectric, and nuclear plants are grouped according to their sizes (that is, economies of scale of plants). This classification is to shed light on future emphasis on the development of large-size power plants, given that large plants are superior to small ones in terms of thermal efficiency, capital costs and environmental impact. Thus there are major potential gains that can be achieved by installing large units in terms of economic and thermal efficiency as well as environmental benefit (see Section 3.4 for a further discussion).

Second, coal-fired power of unit capacity of less than 200 MW represents a worst-case option from the point of view of CO_2 emissions, because the rest of power supply technologies has the potential to reduce a certain amount of CO_2 emissions. Thus, in the present study, it has been chosen as the reference, whereas other options are regarded as abatement technologies. These abatement technologies will each be measured against this reference in terms of both the increased costs and the amount of CO_2 emissions reduced.

The power generation technologies are represented in the power planning model in terms of their technical, economic and environmental characteristics.

8.5   Description of technology by its economic, technical and environmental parameters

The power planning model explicitly lists all the power generation technologies given in Section 8.4, with each characterized by its initial investment, transmission and distribution costs, fixed and variable operating and maintenance costs, generation efficiency, plant utilization, plant lifetime, construction lead times, fuel requirements, dates of introduction, maximum rates of expansion, and emissions coefficients. This section will document the sources of data and present some estimates for a number of parameters required for running the power planning model.

8.5.1   Initial investment

The initial investment cost varies significantly among plants, depending on type, size and state-of-the-art of plants. Investment by plant and size has been taken from various sources.

8.5.1.1 Coal-fired plant
On average, the nationwide investment cost over the period 1989-1991 was 2206 yuan/kW (Ministry of Energy, 1991a, 1992a). On the basis of this, the investment cost is assumed to be 2000 yuan/kW for a coal-fired plant with a capacity of more than 300 MW, 2150 yuan/kW for a plant with a capacity between 200 MW and 300 MW, and 2500 yuan/kW for a plant with a capacity of less than 200 MW.[15] If the installation of flue gas desulphurization takes place, coal-fired plant generation costs are re-evaluated by adding 20-30% to the estimates of both initial investment and fixed operations and maintenance costs following the assumption of Li, Liu and Liu (1990).

8.5.1.2 Nuclear plant
The investment cost for a unit capacity of 600 MW and above is assumed to be 8000 yuan/kW (State Planning Commission, 1993). As for a unit capacity of 300 MW, because of its small size, the corresponding figure is assumed to be 10~15% higher than that of 600 MW (Wen, 1992).

---

[15] Unless specified otherwise, all the investment costs in this chapter are measured at 1990 prices.

8.5.1.3 Hydroelectric plant

The investment cost for a unit capacity of more than 25 MW has been derived from Z. Yang (1992), while the corresponding figure for mini-hydroelectric power has been taken from Zhu *et al*. (1991). The investment cost of *hydroplant of the pumped storage type* is based on the Ming Tombs pumped storage power station under construction.

8.5.1.4 Imported natural gas-fired plant

A natural gas-fired plant offers the advantages of low unit capital cost, low pollutant emissions, and high efficiency. Following the State Planning Commission (1993), the investment cost of a gas-fired plant has been taken as three quarters of the investment cost for a coal-fired plant of the same size. Moreover, apart from capital cost of gas-fired plant itself, the investment in receiving terminal infrastructure, which is to be discussed later, should also be considered.

8.5.1.5 Centralized and decentralized solar PV power plants

The investment costs of these two power generation technologies are assumed to be 24,200 yuan/kW and 35,300 yuan/kW in 2000, while the corresponding figures in 2010 are assumed to be 15,150 yuan/kW and 24,670 yuan/kW. These costs are the average values of the corresponding high and low costs assumed by the State Planning Commission (1993).

8.5.1.6 Wind-driven power plant

The investment cost of a large- and medium-size wind power generator is assumed to be 7160 yuan/kW in 2000 and 6500 yuan/kW in 2010, while the corresponding figure for a *decentralized mini-wind power generator* is assumed to be 10,000 yuan/kW (State Planning Commission, 1993).

8.5.1.7 Biomass-based power plant

Following the State Planning Commission (1993), its initial investment cost is assumed to be 5000 yuan/kW.

8.5.1.8 Geothermal-based power plant

Following the State Planning Commission (1993), its initial investment cost is assumed to be 7000 yuan/kW.

8.5.2 Coal mining investment and delivery cost per kW of capacity

Just as the name implies, coal-fired plants take coal as an input fuel. The annual amount of coal consumed per kW of capacity installed depends on gross coal consumption rate, house service consumption rate and utilization hours of generating unit. Given the fact that the current low coal price does not reflect the real cost of coal production, any cost estimate that takes the coal price instead of the real cost of coal production and delivery into account for comparative analysis of coal-fired plant and other kinds of power plants is relatively cost-biased towards coal plants and hence is considered to be unrealistic.

Our cost estimate is based on the real cost of coal production and delivery. Coal mining investment per kW of capacity installed is derived by multiplying the unit cost of coal production by the annual coal consumption per kW of capacity, while the corresponding delivery cost of coal is calculated by multiplying the unit cost of coal transportation by the annual coal consumption per kW of capacity.

8.5.3 Natural gas receiving terminal investment per kW of capacity

Natural gas receiving terminal investment per kW of capacity installed is derived by multiplying the unit cost of gas receiving terminal by the annual gas consumption per kW of capacity. The former has been taken from the State Planning Commission (1993), while the latter, similar to coal-fired plants, has been determined by gross gas consumption rate, house service consumption rate and utilization hours of generating unit.

8.5.4 Operating and maintenance cost

Annual fixed operating and maintenance (O&M) cost associated with the capacity installed is charged regardless of plant utilization. Fixed O&M cost is assumed to be 3% of the investment cost of fossil fuel-fired plants, 2.7% for nuclear plants, and 1.2% for hydroelectric plants (including hydroelectric pumped storage) (Li *et al.*, 1990). As for other types of plants, the fixed O&M costs are assumed to range from 0.3% to 2.0% of the corresponding investment costs.

Variable annual O&M cost is proportional to production level. It is assumed to be 0.2-0.3 cent/kWh for fossil fuel-fired plants, 0.5 cent/kWh for nuclear plants, and 0.1 cent/kWh for hydroelectric plants (including hydroelectric pumped storage) (Li *et al.*, 1990; Shi *et al.*, 1992).

8.5.5 Transmission and distribution cost

As nearly 70% of the total exploitable potential of hydropower is located in the southwest of China remote from power load centres, the highest transmission and distribution (T&D) investment cost is assumed for hydroelectric power. In contrast, plants such as nuclear power, pumped storage power, and imported natural gas-fired power are assumed to be built near to the load centres, thus requiring the lowest T&D investment cost. The T&D cost for coal power lies somewhere in between. The assumption about the T&D cost by plant is given in Table 8.1.

8.5.6 Gross fuel consumption of thermal power generation

Gross fuel consumption of thermal power generation is referred to as the amount of fuel consumption per unit of electricity generation. It is usually measured in gce/kWh and serves as one of the main techno-economic indicators of power systems. The larger the size of the power plant, the higher the generation efficiency and hence the less will be the corresponding CO_2 emissions. In China, the average gross fuel consumption of thermal power generation in 1990 was 392 gce/kWh corresponding to an average generation efficiency of 31.4%.[16] It is conceivable that with the increases in plant scale in the future the average fuel consumption will go down.

Table 8.1   *Transmission and distribution cost of power plant*

Generation method	Transmission and distribution cost
Coal power (<200 MW)	500 yuan/kW
Coal power (≥200 MW)	700 yuan/kW
Nuclear power	500 yuan/kW
Hydroelectric power (>25MW)	1500 yuan/Kw
Pumped storage power	500 yuan/kW
Imported natural gas power	500 yuan/kW

Sources: State Planning Commission (1993); own estimates.

Table 8.2 shows the assumed gross fuel consumption of thermal power plants in 2000 and in 2010 and the corresponding CO_2 emission index. In determining its future value, it is assumed that gross fuel consumption of coal power will go down in order to reflect ongoing technological development.

Table 8.2   *Gross fuel consumption and CO_2 emission index of thermal power*

Generation method	Gross fuel consumption		CO_2 emission index[a]	
	2000	2010	2000	2010
Coal power (200 MW)	390	365	100	100
Coal power (300 MW)	345	340	88	93
Coal power (600 MW)	320	315	82	86
Natural gas power	246	246	39	42

[a] Coal power (200 MW) = 100.

Sources: Li, Liu and Liu (1990); Energy Research Institute (1991); Williams (1992); own estimates.

---

[16] The generation efficiency of thermal power is defined as a ratio of 123 gce/kWh to its gross fuel consumption.

It can been observed that the CO_2 emission index for natural gas power is even less than half that of coal-fired plants. This is partly because natural gas-fired plants offer greater efficiency than coal-fired plants and partly because carbon emissions from natural gas combustion are the lowest of all fossil fuels when measured in tonnes of carbon per unit of energy content (see Table 3.15).

8.5.7 Plant lifetime

The assumed lifetime differs among plants. *Ceteris paribus*, the longer the lifetime, the lower the generation cost per kWh, as the total fixed costs can be distributed over a larger power output. As is usual in comparative analysis of coal-fired power and nuclear power (OECD, 1989), a lifetime of thirty years is assumed for both fossil fuel-fired and nuclear plants, although no nuclear plants with a capacity of greater than 200 MW have operated that long (Virdis and Rieber, 1991). As for hydroelectric power, a lifetime is assumed to be fifty years. As for wind-based power generation, solar PV power, biomass-based power generation and geothermal-based power generation plants, their lifetimes are all assumed to be thirty years.

8.5.8 Construction lead times

Lead time is defined to be the number of years from ordering to putting into operation. The more capital-intensive the power plant, the more sensitive its cost to the length of time prior to operation. Given the discount rate, investment expenditures capitalized over a longer lead time incur higher investment costs.

Lead times vary widely, depending on both plant type and size. The experience has shown that lengthening of average lead times is greater for nuclear plants than for coal plants (OECD, 1985). Thus lead times for coal plants are taken as 3-6 years by comparison with 7-8 years for nuclear plants (Li, Liu and Liu, 1990). As for hydropower, the spread of lead time is from three years for mini-hydropower to four and a half years for pumped storage power and to ten years for large size plants.

8.5.9 Plant unavailability

Plant unavailability is defined to be one minus annual availability. The parameter relates to electricity production and plant maintenance, and plays a role in reducing fuel input consumption. This unavailability is divided into two parts: scheduled and forced outages. The annual scheduled outage is assumed to be 60 days for coal plants with a capacity of more than 300 MW, 53 days for plants with a capacity between 200 MW and 300 MW, and 45 days for plants with a capacity of less than 200 MW. The corresponding fraction of unavailability due to forced outage is assumed to be 12.77%, 7% and 6% respectively. These two figures are assumed to be 40 days and 5% for gas-fired plants, and 60 days and 14.36% for nuclear plants. By comparison with fossil-fired plants, hydropower plants are of low unavailability and forced outage. The corresponding figures are assumed to be 20-30 days and 1-2%.[17]

8.5.10 Other technical parameters

Other technical parameters include house service consumption rate (HSCR) of various power plants, transmission and distribution efficiency of the power grid, thermal efficiency of nuclear power, and unit consumption of nuclear fuel.

The HSCR is referred to as the amount of electricity consumed by station auxiliaries and is measured by a percentage of gross electricity generation. The assumed HSCR by thermal, hydropower and nuclear plants is given in Table 8.3. The corresponding figure for other HSCR is negligible.

Table 8.3   *House service consumption rate of conventional power plant*

Generation method	House service consumption rate
Coal power (<200 MW)	8.00%
Coal power (≥ 200 MW)	7.50%
Hydroelectric power	0.25%
Pumped storage power	0.30%
Nuclear power	4.00%
Natural gas power	3.00%

Sources: Provided by Prof. Li Weizheng; Zhao (1989); own estimates.

Transmission and distribution efficiency of the power grid is defined to be one minus line losses in transmission and distribution. Currently, the line losses are over 8% of the total electricity supplied (State Statistical Bureau, 1992b). With the upgrading of the power grid, the figure is expected to go down to 8% in 2000 and 7% in 2010 (Zhang, 1991).

As for thermal efficiency of nuclear power, 33% is assumed for nuclear power plants with a unit capacity of 300 MW and 35% for plants with a unit capacity of 600 MW and above. The unit consumption of nuclear fuel is assumed to be 33000 MW·day per ton of uranium (Shi *et al.*, 1992).

8.5.11 CO_2 emission coefficients

The amount of CO_2 emissions per unit of energy, the so-called CO_2 emission coefficients, vary among fuels. By convention, the emission coefficients are expressed in terms of the weight of the carbon atom only, not the weight of the

---

[17] The data in this section are based on those used for the Electric Power Planning in Guangdong Province and are provided by Prof. Li Weizheng.

entire CO_2 molecule (Edmonds and Darmstadter, 1990). The coefficients used for calculating CO_2 emissions associated with energy consumption are shown in Table 3.15, which are generally considered suitable for China.

8.6   A comparison of alternative power plants

Before addressing their most cost-efficient mix, in this section I will perform a comparison of alternative power plants in terms of both the generation cost and the marginal cost of carbon abatement. Its purpose is to rank the power plants examined in terms of their cost-effectiveness and to shed light on the priorities of abatement investments, the latter being viewed as a precondition for successful joint implementation projects.

I start with discussing the generation costs of power plants of various types. The International Union of Producers and Distributors of Electrical Energy (UNIPEDE) recommends that electricity generation cost should be expressed in constant money units levelized over the lifetime of the plant. This cost is the so-called levelized cost.[18] It is made up of three parts - capital recovery cost, operation and maintenance cost, and fuel cost:

$$LC_i = CRC_i + OMC_i + FC_i$$

where

CRC_i	=	Capital recovery cost of power plant i.
FC_i	=	Annual fuel cost of power plant i.
LC_i	=	Levelized cost of power plant i.
OMC_i	=	Annual operation and maintenance cost of power plant i.

CRC_i is an annuity. Assuming that during the construction lead time the same amount of investment expenditures is spent per annum, CRC_i is then determined by spreading capital cost (including interest charges) over total output during the lifetime of power plant i:

---

[18] Levelized cost spreads total generation cost over total output to arrive at a figure which, if charged for each kWh, would exactly balance costs and income (OECD, 1989). Initially it was put forward by UNIPEDE to compare cost of nuclear power generation with that of conventional power generation. Afterwards the levelized cost methodology is accepted and widely used by the OECD and the IAEA (International Atomic Energy Agency) nations.

$$CRC_i = \left(\frac{cc_i}{cl_i \cdot u_i} \right) \left(\frac{(1+\alpha)^{cl_i}-1}{\alpha} \right) \left(\sum_{t=1}^{t=l_i} \frac{1}{(1+\alpha)^t} \right)^{-1}$$

where

cc_i	=	Capital cost per incremental unit of capacity of power plant i.
cl_i	=	Construction lead time of power plant i.
l_i	=	Lifetime of power plant i.
u_i	=	Average utilization of power plant i (hours per year).
α	=	Annual discount rate.

Let us now turn to the marginal cost of carbon abatement. For a given level of service, each advanced plant will have high capital costs but can save a certain amount of energy with respect to the reference. Associated with those savings in fuel is a reduction in CO_2 emissions. Dividing the increased cost of an advanced plant by the corresponding amount of CO_2 emissions reduced, the incremental cost per unit of CO_2 reduction, that is, marginal cost of CO_2 reduction by an advanced plant, can be expressed as follows:

$$MC_i = - \frac{LC_i - LC_j}{ECO2_i - ECO2_j} \qquad j=Coal \ power \ (<200MW), \ i \neq j$$

where

$ECO2_i$	=	CO_2 emissions per unit of electricity generation by plant i.
MC_i	=	Marginal cost of CO_2 reduction by advanced plant i.

Note that in calculating the marginal cost of carbon abatement, I use the levelized cost rather than the capital cost. The reason for such a choice is the following: given that shortages of capital resources are often the main constraint on the adoption of new technologies, the marginal cost of carbon abatement calculated on a basis of capital cost could be useful for policymakers as a means of ranking alternative technologies. Its limitations are also obvious, however, because no account is taken of discounting costs over time, operation and maintenance cost, and fuel cost. This could result in a bias towards conventional plants, because advanced plants, though costly, are more energy-efficient compared with the reference.

Table 8.4 shows both the generation cost and the marginal cost of CO_2 reduction by each plant at a 10% discount rate. They clearly indicate the cost-effectiveness of each plant. The lower the levelized cost, the more cost-effective the plant. This means that renewable energy plants, such as wind and PV plants, are still too costly in comparison with conventional coal and hydroelectric plants.

Technically, they have yet to be proved for large-scale electricity production. Nuclear power, though more competitive than renewable power, is less competitive than coal power. In order to ease the increasing energy shortages and because nuclear power provides, with the exception of hydroelectric power, the only so far proven method for enormous potential for large-scale generation of electricity without a directly parallel production of CO_2 emissions, however, nuclear power's share in total electricity generation is expected to rise. This rise would come at higher electricity prices than would otherwise have been the case, given an inverse relationship between electricity prices the percentage of electric power generated from nuclear power. Imported gas-fired power is even less economical than nuclear power, although it is just the other way around in the Western countries (MacKerron, 1992). Thus, the development of gas power is only a limited solution to meet the power needs in the energy-deficient but economically developed coastal areas.

It can be observed that large coal-fired plants (200 MW and above) and hydroelectric plants have negative marginal costs of CO_2 reduction in comparison with the reference. Let us have a closer look at the interesting results.

From an economic point of view, that the marginal costs of carbon abatement by power plants are negative implies that these plants are less expensive than the reference. For large coal plants, the interpretation is straightforward. Large units, in comparison with the reference, are energy-efficient and, because of their economies of scale of plants, are also cost-effective. Thus, it is conceivable that the marginal costs of CO_2 reduction by large units are negative. This conclusion remains unchanged no matter what the coal price and the level of discount rates are.

Now that large plants are superior to small ones in terms of thermal efficiency, capital costs and environmental impact, why has their share in total capacity installed so far been low? This is partly due to the lack of manufacturing capacity for large units. At present, domestic capacity is only able to manufacture one unit of 600 MW and 12-13 units of 300 MW (Hu *et al.*, 1992). It can only meet about 25% of the newly-added thermal capacity required per annum during the coming decade. The second reason is because of the difficulties in mobilizing the necessary large investment resources. A disturbing consequence is that some local governments under pressure of acute power shortages are continuing to invest in large numbers of small coal-fired plants, although national policy emphasizes development of 300 MW and 600 MW units (World Bank, 1994). The third reason is related to the disappointing performance of domestically-produced large units.[19] In comparison with small plants, the equivalent availability factor of large plants is low. This is partly because large units have long outages (see Section 8.5.9) and partly because a number of large units are still unable to operate stably at the nominal outputs designed (the full capacity).

---

[19] This was also experienced in adopting 500 MW generating units in the United Kingdom (see Section 3.4).

Table 8.4 *A comparison of alternative power plants at a 10% discount rate*

	Capital recovery cost (cent/kWh)[a]	O&M cost (cent/kWh)[a]	Fuel cost (cent/kWh)[a]	Levelized cost (cent/kWh)[a]	Marginal cost[b] (yuan/tC)[a]
Coal power (<200 MW)	5.320	1.805	11.870	18.995	–
Coal power (200 MW ~ 300 MW)	4.811	1.561	10.446	16.818	-744.320
Coal power (>300 MW)	4.960	1.603	9.685	16.248	-602.083
Hydroelectric power (>25 MW)	13.084	1.657	–	14.740	-167.558
Mini-hydroelectric power (≤25 MW)	15.605	1.865	–	17.471	-60.025
Pumped storage hydroelectric power	21.606	2.671	–	24.277	208.045
Nuclear power (300 MW)	21.086	5.867	2.391	29.345	407.627
Nuclear power (600 MW ~ 1000 MW)	20.219	5.646	2.255	28.119	359.367
Imported natural gas-fired power	10.581	1.497	21.101	33.179	917.289
Wind-driven power generation	28.065	4.969	–	33.034	552.926
Decentralized mini-wind power generator	39.040	8.361	–	47.401	1118.777
Centralized solar PV power	112.105	6.284	–	118.389	3914.607
Decentralized solar PV power	162.870	4.606	–	167.476	5847.905
Biomass-based power generation	26.520	4.750	–	31.270	483.448
Geothermal-based power generation	16.386	8.723	–	25.109	240.807

[a] Measured at 1990 prices.
[b] Marginal cost of carbon abatement.

Source: Own calculations.

In comparison with coal plants, the results for hydroelectric plants need to be interpreted with caution. This is because their comparisons with the reference depend greatly on coal price and level of discount rates. For the present study, as pointed out in Section 8.5.2, I use the real user price of coal, which reflects the real cost of coal production and delivery. If the subsidized price of domestic coal is used, hydroelectric plants are not yet competitive with the reference. This is the reason for the current underdevelopment of hydroelectric power. Moreover, because hydroelectric plants are capital-intensive and have long construction lead times in comparison with small coal-fired units, the level of discount rates does matter. Given 10% as the cost of capital, hydropower electricity would be cheaper than coal-fired electricity. Should the cost of capital be 13%, however, hydropower electricity would become costlier than coal-fired electricity.

8.7   The business-as-usual scenario for electricity supply

As a starting point for the reduction scenarios, a business-as-usual (or baseline) scenario needs to be developed for electricity supply in China over the period to 2010. The scenario assumes no policy intervention directly aimed to limit the rate of CO_2 emissions from the electricity supply. The quantity and type of an additional new plant needed over time are determined optimally according to the relative economics of the alternative options subject to constraints on electricity demand and the rate of plant construction within specific time periods. The results for the baseline scenario are given in Tables 8.5 and 8.6.

The baseline scenario is characterized by a rapid growth of the power industry during the period under consideration. As shown in Table 8.5, total generating capacity is expected to rise from 137.94 GW in 1990 to 275.70 GW in 2000 and to 539.63 GW in 2010, with an average annual net add-up of 14 GW from 1990 to 2000 and 26 GW thereafter to 2010.[20] Accordingly, there will be an increasing demand for investment in electric power in order to materialize such add-up. Measured as a percentage of GNP, capital investment in the electricity sector is estimated to rise from 1.89% in 1990 to 2.19% in 2000. With the subsequent slowdown of economic growth and hence electricity demand, capital investment in the electricity sector as a share of GNP is calculated to go down slightly to 2.14% in 2010, although an increasing expansion of capital-intensive plants, such as nuclear power and hydroelectric power that have higher unit investment costs than coal power, will turn out to have the opposite effect.

---

[20] Although an average annual growth rate of generating capacity is slightly lower during the second period than during the first one, its average annual add-up during the second period is almost two times that during the first period. This is mainly because total generating capacity in 2000 is almost two times that in 1990.

Table 8.5   *Main results for the baseline scenario*

	1990	2000	2010
Total generating capacity installed (GW)	137.94	275.70	539.63
Coal used for electricity generation (Mt)	272.04	543.18	976.96
As a share of total coal consumption (%)	25.78	34.40	40.40
Gross coal consumption (gce/kWh)	398[a]	352.97	338.78
Capital investment in the electricity sector			
as a share of GNP (%)	1.89	2.19	2.14
CO_2 emissions from coal-fired power (MtC)	126.50	252.58	454.29

[a] Capacity installed of 6000 kW and above.

Sources: For data of 1990, State Statistical Bureau (1992a, 1992b); others from own calculations.

A breakdown of total generating capacity by plant is shown in Table 8.6. It can be seen that coal-fired power plants still predominate, although the role of alternatives is growing. Accordingly, the amount of coal consumed for electricity generation grows rapidly. Its share in total coal consumption rises from 25.8% in 1990 to 34.4% in 2000 and to 40.4% in 2010. Correspondingly, this will lead to an increase in emissions of SO_2 and CO_2 within the power sector itself, although the decreasing direct use of coal will alleviate the environmental impacts of coal use as a whole.[21] The calculations show that the related CO_2 emissions from coal-fired plants are expected to rise from 126.5 MtC in 1990 to 252.6 MtC in 2000 and to 454.3 MtC in 2010, with an average annual growth rate of 7.2% from 1990 to 2000 and 6.0% thereafter to 2010. The slowdown of CO_2 emissions during the second period is largely due to a decreasing economic growth and, as shown in Table 8.6, an increasing penetration of efficient or carbon-free plants.

With respect to the scale of coal plants, as shown in Table 8.6, more large units are expected to be put into operation during the period under consideration compared with the current composition of plants, although a variety of problems as discussed in Section 8.6 need to be resolved in practice in order to promote the rapid moving up to large units.[22] The results are generally considered in

---

[21] There are at least two reasons for this. First, boilers used in power plants are much more efficient than coal stoves used by households. Second, control of pollution caused by point sources like power plants is less costly than that of nonpoint sources like the nationally widely-used coal stoves.

[22] Although it goes beyond the scope of this study, it should be pointed out that, in constructing large coal-fired power plants, the availability of water resources and the quality of these resources after use should also be considered. Similarly, the ecological damages from constructing large hydroelectric power plants must be considered.

accordance with the government focus on increasing plant scales. Because large units are more energy-efficient than small units, the moving up to large units will lead to a decline in gross coal consumption of power plants. As shown in Table 8.5, the average coal consumption of power plants is estimated to go down from 398 gce/kWh in 1990 to 353 gce/kWh in 2000 and to 339 gce/kWh in 2010. Accordingly, the average generation efficiency will go up from 30.9% in 1990 to 34.8% in 2000 and to 36.3% in 2010.

As for nuclear power, although there is little prospect of dramatic increases during the period under consideration, nuclear power begins to make a useful contribution to China's electricity supply. As shown in Table 8.6, the share of nuclear power stations in total generating capacity is expected to rise to 0.76% in 2000 and to 3.45% in 2010, whereas in 1990 no nuclear power stations were commissioned. This suggests that there will be a rapid development of nuclear power after its initial development stage to the year 2000.

8.8   Carbon abatement: the cost-efficient mix of power plants

As discussed in Section 7.3.3, the price of coal in 2010 will rise by 65% as a result of the imposition of a carbon tax of 205 yuan per ton of carbon in order to achieve a 20% cut in CO_2 emissions in 2010. In this section, I will examine the impacts of such a change on the mix of power plants. In doing so, it is assumed that the price of coal changes linearly over time, implying that the price of coal in 2000 will have risen by 32.5%.

Increasing the coal price makes coal less attractive as a fuel. Thus, it seems reasonable to assume that this will induce some shifts towards low-carbon or carbon-free plants. But in the baseline scenario, because the levelized cost of hydropower is lower than that of coal power, hydroelectric power has already been implemented to its maximum potential (that is, upper bound). Thus, there is no room left for the substitution of hydroelectric power for coal power if the bound on hydroelectric power remains unchanged. This also applies to large coal power with a unit capacity of more than 300 MW, which has also been implemented to its maximum potential in the baseline case. As for nuclear power, even in the case where the carbon tax is imposed, the levelized cost of nuclear power is still higher than that of coal power. Thus, there will be no switching to nuclear power.

Facing the typical corner solution of the linear programming, which begins with the least-cost technology and progresses to others also in the least-cost sequence, I have investigated three alternative policy simulations:

Simulation 1: Moving up to large coal-fired units (>300 MW);
Simulation 2: Switching to hydroelectric power (>25 MW); and
Simulation 3: Switching to nuclear power (600~1000 MW).

Table 8.6 Generating capacity installed for the baseline scenario over the period 1990-2010

	1990		2000		2010	
	GW	%	GW	%	GW	%
Coal power (<200 MW)	61.94	44.90	47.34	17.17	32.80	6.08
Coal power (200 MW ~ 300 MW)	32.70	23.70	127.55	46.26	281.95	52.25
Coal power (>300 MW)	6.80	4.93	25.20	9.14	60.20	11.16
Hydroelectric power (>25 MW)	22.38	16.22	50.00	18.14	105.00	19.46
Mini-hydroelectric power (≤25 MW)	13.67	9.91	17.80	6.46	28.50	5.28
Pumped storage hydroelectric power	0.03	0.02	4.00	1.45	8.00	1.48
Nuclear power (300 MW)	-		0.30	0.11	0.60	0.11
Nuclear power (600 MW ~ 1000 MW)	-		1.80	0.65	18.00	3.34
Imported natural gas-fired power	0.40	0.29	0.80	0.29	2.40	0.44
Wind power	0.02	0.01	0.78	0.28	1.55	0.29
Others	0.03	0.02	0.13	0.04	0.63	0.12
Total	137.94	100.00	275.70	99.99	539.63	100.00

Source: Own calculations.

Table 8.7 shows the results of the three simulations. Although the share of coal-fired power plants in total generating capacity installed remains unchanged by comparison with the baseline, the average gross coal consumption of coal-fired plants in Simulation 1 is expected to go down to 333.46 gce/kWh in 2010 because of the accelerated replacement of small units by large efficient ones. Consequently, this will lead to a reduction in CO_2 emissions by 1.6% in 2010. Moreover, such a replacement requires increased investment in electric power. Measured as a percentage of GNP, capital investment in the electricity sector in 2010 is estimated to rise from 2.14% in the baseline scenario to 2.25% in Simulation 1, although total generating capacity installed remains unchanged in comparison with the baseline.

Compared with Simulation 1, switching to hydroelectric power in Simulation 2 will also bring the average gross coal consumption of coal-fired plants down, but to a lesser extent. Because hydroelectric power is carbon-free energy, however, switching away from coal-fired power towards hydroelectric power will lead to a larger reduction in CO_2 emissions than in Simulation 1. In addition, because hydroelectric power operates at a smaller load factor than coal power, in order to meet the same electricity demand as Simulation 1, total generating capacity installed in Simulation 2 is thus larger than in Simulation 1. Combined with hydroelectric power being more capital-intensive than coal power, we expect that more capital investment is required in Simulation 2. This is confirmed in Table 8.7, which shows that capital investment in the electricity sector as a share of GNP in 2010 will go up from 2.25% in Simulation 1 to 2.38% in Simulation 2.

In 2010, Simulation 3 yields almost three times the amount of CO_2 reductions of Simulation 1, but to a lesser extent than in Simulation 2. This is mainly because nuclear power is even more capital-intensive than hydroelectric power so that switching away from coal-fired power towards nuclear power is less than that towards hydroelectric power in Simulation 2.

Consequently, the average gross coal consumption of coal-fired plants in Simulation 3 will go down less than in Simulation 2. In terms of capital investment in the electricity sector as a share of GNP, although nuclear power is more capital-intensive than hydroelectric power, the figure for Simulation 3 is still smaller than that for Simulation 2, since nuclear power operates at a much higher load factor than hydroelectric power. But compared with Simulation 1, mainly because nuclear power is much more capital-intensive than coal power, the corresponding share for Simulation 3 is larger, although total generating capacity installed remains almost unchanged in the two simulations.

Using the CGE model that has been discussed in Chapters 5 to 7, the GNP effects of the three alternative policy simulations can be calculated. Table 8.7 shows the results for 2010. Since more capital is diverted to the electricity sector of capital-intensive nature, we expect that the GNP growth will decline. Moreover, the greater the extent to which such a shift of capital takes place, the larger the GNP losses. This is confirmed in Table 8.7, which shows that, relative to Scenario 1 in Section 7.3.2, GNP falls by 0.92% in 2010 in Simulation 1, 1.58% in Simulation 2, and 1.38% in Simulation 3. Besides, the accelerated replacement

Table 8.7 *Main results for the increase in real coal price by 65% in 2010*

	Large coal power		Hydroelectric power		Nuclear power	
	2000	2010	2000	2010	2000	2010
Reduction in CO_2 emissions relative to the baseline	-1.3%	-1.6%	-3.1%	-5.7%	-2.5%	-4.7%
Capital investment in the electricity sector as a share of GNP (%)	2.34	2.25	2.38	2.38	2.35	2.34
Gross coal consumption (gce/kWh)	348.37	333.46	351.58	336.96	352.19	337.64
Total generating capacity installed (GW)	275.69	539.63	277.21	545.01	275.32	538.27
Share of coal power in total capacity installed (%)	72.86	69.93	70.51	65.70	71.31	67.06
Share of hydropower in total capacity installed (%)	26.04	26.22	28.40	30.49	26.08	26.29
Share of nuclear power in total capacity installed (%)	0.76	3.45	0.76	3.41	2.28	6.24
GNP relative to the baseline (-: declines)	-1.535%		-1.546%		-1.542%	
GNP relative to Scenario 1 in Section 7.3.2 (-: declines)	-0.92%		-1.58%		-1.38%	
Price of electricity relative to the baseline (+: increases)	+23.19%		+23.49%		+23.40%	
Price of electricity relative to Scenario 1 in Section 7.3.2 (+: increases)	+0.33%		+0.60%		+0.52%	

Source: Own calculations.

of small coal-fired units by large efficient units, or by capital-intensive plants, such as nuclear power and hydroelectric power will lead to a rise in the price of electricity. According to the calculations with the CGE model, however, the price of electricity rises, but to a limited extent relative to Scenario 1 in Section 7.3.2.

8.9   Summary and conclusions

Unlike Chapter 7, this chapter aims to shed light on technological aspects of carbon abatement in the electricity sector. Thus, it is devoted to satisfying electricity planning requirements in the CO_2 context. There were at least four reasons for studying the electricity sector. First, the electricity sector is a major consumer of primary energy. Second, there are many more interfuel substitution possibilities in the sector compared with other sectors. Third, the results obtained from studying the electricity sector can contribute to forming a necessary basis for China's development of joint implementation projects with other countries, because projects for both increased generation efficiency and interfuel substitution in the sector are development priorities of the Chinese government. Fourth, time constraints did not allow studying the Chinese energy system as a whole, although it would have been desirable.

For a cost-effective analysis of carbon abatement options in China's electricity sector, a technology-oriented optimization model for power system expansion planning has been developed. This model has been adapted from the MARKAL model. It chooses the minimization of discounted cost over the entire planning horizon as its objective function and incorporates a number of power-related constraints adopted by MARKAL. In the power planning model that has been programmed in GAMS, 15 types of power plants are represented in terms of their technical, economic and environmental parameters. Specifically, these parameters include initial investment, transmission and distribution costs, fixed and variable operating and maintenance costs, generation efficiency, plant utilization, plant lifetime, construction lead times, fuel requirements, dates of introduction, maximum rates of expansion, and emissions coefficients. The model allows for substitution from high-carbon fossil fuels and technologies towards low-carbon and carbon-free counterparts and for interactions between periods to cope with carbon limits.

In order to examine the cost-effectiveness of the 15 types of power plants considered and hence to shed light on the priorities of abatement investments, the latter being viewed as a precondition for successful joint implementation projects, a comparison of the power plants has been made in terms of both the levelized cost of generation and the marginal cost of CO_2 reduction at a 10% discount rate. Such a comparison shows that renewable energy plants, such as wind and PV plants, are still too costly by comparison with conventional coal and hydroelectric plants, although they can generate electricity without a directly parallel production of CO_2 emissions. It has also been found that large coal-fired plants (200 MW and above) and hydroelectric plants have negative marginal costs of CO_2

reduction in comparison with small coal-fired plants (<200 MW), the latter being chosen as the reference since they represent a worst-case option from the point of view of CO_2 emissions (see Tables 8.2 and 8.4). Thus, these plants should be given priority in carbon abatement investments. In practice, however, in order to promote the moving up to large units, efforts should be directed towards resolving a variety of problems, including the lack of domestic capacity for manufacturing large units, the difficulties in mobilizing the necessary large investment resources, and the disappointing performance of domestically-produced large units.

Using the power planning model and based on the macroeconomic results obtained through the CGE model concerning GNP growth and energy demand, a baseline scenario for electricity supply over the period to 2010 has been developed. The baseline scenario is characterized by a rapid growth of the power industry, with an average annual net add-up of 14 GW from 1990 to 2000 and 26 GW thereafter to 2010. Accordingly, capital investment in the electricity sector as a share of GNP is calculated to go up from the current level. With respect to a breakdown of total generating capacity by plant, the calculations show that coal-fired power plants still predominate, although the role of alternatives is growing. Accordingly, the amount of coal consumed for electricity generation grows rapidly, thus increasing its share in total coal consumption. This will lead to an increase in CO_2 emissions within the power sector itself, although the decreasing direct use of coal will alleviate the environmental impacts of coal use as a whole. With respect to the scale of coal plants, it has been shown that more large units are expected to be put into operation during the period under consideration compared with the current composition of plants. This will bring the average gross coal consumption of coal-fired plants down. Besides, the calculations show that nuclear power begins to make a useful contribution to China's electricity supply, although there is little prospect of dramatic increases until the year 2010.

In Section 8.8, the impacts have been examined of a 65% rise in the price of coal as a result of the imposition of a carbon tax of 205 yuan per ton of carbon in order to achieve a 20% cut in CO_2 emissions in 2010. In this regard, three alternative policy simulations have been investigated, all of which are relative to the baseline. Simulation 1 shows the accelerated moving up to large coal-fired units, Simulation 2 the accelerated expansion of hydroelectric power, and Simulation 3 the accelerated expansion of nuclear power. Of the three simulations, the calculations for Simulation 1 suggest the largest reduction in average gross coal consumption of coal-fired plants, the smallest increase in capital investment, but the smallest reduction in CO_2 emissions. The calculations for Simulation 2 suggest the highest increase in capital investment, but the largest reduction in CO_2 emissions. The calculations for Simulation 3 suggest that capital investment and reduction in CO_2 emissions lie in between Simulations 1 and 2, but are closer to those for Simulation 2. With respect to the effects on the GNP losses and the price of electricity, the calculations through the CGE model show that there are slight differences among the three alternative policy simulations. This, combined with the effectiveness in a reduction in CO_2 emissions, would suggest the accelerated expansion of hydroelectric power and nuclear power in the carbon abatement investments (of course, the safety and radioactive waste disposal of nuclear power plants and the ecological damages from constructing large hydroelectric power plants must be taken into account). This finding is in line with the government investment priority regarding the electricity sector, which has been set with aims to reduce pressure on transportation and air pollution but without considering the greenhouse effect. This implies that the development of hydroelectric power and nuclear power will be accelerated, if curbing CO_2 emissions is taken into account.

9 Conclusions

This study is the first systematic and comprehensive attempt to deal with the economic implications of carbon abatement for the Chinese economy in the light of the economics of climate change. This chapter summarizes the main conclusions. Given the purpose of the study, the conclusions focus on the following topics:

- Analysis of the Chinese energy system: implications for future CO_2 emissions;
- Macroeconomic analysis of CO_2 emission limits for China; and
- Cost-effective analysis of carbon abatement options in China's electricity sector.

In addition, relevance and potential use of the results from this study for science and policy-making are discussed. Finally, some suggestions for further methodological and empirical work are provided in order to enrich the policy relevance of the study.

9.1 Analysis of the Chinese energy system: implications for future CO_2 emissions

Chapter 3 centered upon the Chinese energy system and its implications for the nation's future CO_2 emissions. The following conclusions can be drawn from examining China's energy resources and their development, the Chinese energy consumption patterns, the achievements and remaining problems of electricity generation in China, China's energy conservation in an international perspective, historical CO_2 emissions in China, and environmental challenges for the Chinese energy system.

Beginning with the reserves and utilization rates of China's fossil fuels, it is clear that, because of the low level of exploration for oil and natural gas, the proven recoverable coal reserves in China are 12.7 times its proven recoverable oil and natural gas reserves combined (see Table 3.1). As for hydropower, the economically exploitable capacity in China is estimated at as much as 378 GW, which is the largest in the world. Because of the lack of investment and unfavourable exploitation conditions as well as some concerns about environmental impacts of large dams, however, hydropower has so far been underdeveloped. The development of nuclear power in China has just begun. Although the domestic uranium supply can meet the need for short-term nuclear power development, it cannot do so for a long-term large-scale development programme if nuclear power stations are to be based on the currently-used PWRs. With regard to renewables, their role in the energy balance is currently negligible, and there is little prospect of a dramatic increase because renewable systems are not yet competitive with conventional energy supply. Since the country has no readily available substitutes, China is thus bound to rely mainly on coal to fuel the development of its economy and thereby improve the Chinese standard of living in the foreseeable future.

With respect to the Chinese energy consumption patterns, the characteristics can be summarized as follows: main reliance on domestic energy resources; coal-dominant structure of energy consumption; uneven geographical distribution of energy resources and economy; low per capita energy consumption but high energy use per GNP; heavy reliance on biomass energy by rural households; and industry-dominant composition of final energy consumption.

As far as electricity generation is concerned, China has made great strides over the past decade, with an average annual addition of 10 GW. There were also two nuclear power stations successfully put into operation, with a total capacity of 2.1 GW. This marked the end of an era without nuclear power in China. Despite the great achievements in electricity generation, China remains a country with a low penetration of electricity in total final energy consumption. Its power industry still faces some problems, including low unit capacity, underdevelopment of hydropower, the small share of cogeneration units, and deficiencies of capital investment in transmission lines and distribution networks.

With respect to energy conservation, China has successfully moved towards decoupling its GNP growth from energy consumption, with an annual growth of 9.9% for the former but 5.2% for the latter during the period 1980-95. This achievement corresponds to an income elasticity of energy consumption of 0.52, an accumulated energy savings of 630 Mtce and to an annual saving rate of 4.3%. While China has enjoyed such a great success, its energy use per unit of GNP is still among the highest in the world. This exceptionally high energy intensity can be explained by an unusually large share of energy-intensive industrial production in the Chinese economy, a large share of energy-intensive manufacturing in China's industry, a high proportion of coal consumption, and the undervaluation of China's GNP. Thus, direct cross-country comparison of energy use per unit of output value by no means should be interpreted as representing the actual potential of energy conservation. Indeed, a sector-by-sector comparison of energy intensity in physical terms has shown that China's actual potential of energy conservation is much smaller than that from direct comparison of energy use with other countries (see Section 3.5.2). Moreover, in order to encourage future energy conservation investment and hence to materialize such a potential, current subsidies for energy consumption in China should be abolished. Also, consideration may be given to appropriate control over the growth of China's energy supply in order to put pressure of energy conservation on the demand side.

While taking such drastic domestic efforts towards energy efficiency, China badly needs assistance and economic and technical cooperation with the industrialized countries, because of the increasingly large amounts of capital and technical expertise required for energy efficiency gains. Actually, China has already reached such a consensus and therefore has made great efforts to reducing barriers to trade and protecting intellectual property rights in order to facilitate the transfer of advanced energy efficiency and pollution control technologies from abroad. Indeed, it is also in the direct interest of the industrialized countries to encourage such a take-up, since it will act as a relief of pressure on these countries for yet more stringent measures to reduce CO_2 emissions (see Section 7.5.2). This, combined with the commitment made by the industrialized countries at the United Nations Conference on Environment and Development (also called the Earth Summit) in Rio de Janeiro to assist developing countries in reducing their emissions of greenhouse gases would appear to encourage China's request for international assistance.

Past evidence shows, however, that multilateral and bilateral development assistance agencies from the industrialized countries as well as international development banks tend to finance large-scale supply side projects rather than a number of small-scale demand side undertakings aimed at promoting end-use energy efficiency improvements (cf. Kats, 1990, 1992; Chandler *et al.*, 1993). Given the decentralized nature of efficiency projects, the preference for large-scale projects is partly for the sake of reducing administrative overheads. The second reason for the lack of funding for energy efficiency is that some development assistance, especially from bilateral aid agencies, is tied aid, with an official grant or loan offered on the condition that the recipient countries are required to procure goods and services from the donor countries.[1] Clearly, such aid and loan practices are intended to serve the dual purpose of both providing assistance and expanding exports from the donor countries. Kats (1990, 1992) has argued that this financial assistance pattern is innately biased against energy programmes for promoting energy efficiency improvements in developing countries because it results in a preference for large-scale projects that are capital intensive, highly dependent on donor technologies and require major imports in areas of particular export interest to the donor countries (Jepma, 1994). Thus, it is conceivable that developing countries, including China, cannot obtain the most required funds for this purpose from their industrialized counterparts, if this pattern continues. In such a case, their achievements in energy conservation will depend entirely on the extent to which domestic efforts are taken and domestic funds are made available for such efforts.

In order to ease the financial constraint and promote the transfers of technology and know-how, third-party financing (TPF), particularly combined with international funding, has been suggested as one of the financing mechanisms avail-

---

[1] See Jepma (1994) for the definitions of tied aid and its impacts on both donor and recipient countries.

able.[2] By comparison with build-operate-own-transfer (BOOT) that is concerned with investment for expanding energy supply, TPF is concerned with energy conservation of existing energy users (Shunker *et al.*, 1992). Thus, this mechanism is considered a useful means of encouraging the dissemination of energy conservation schemes. Moreover, insofar as the bulk of infrastructure and capital equipment in developing countries has still to be put in place in the course of their industrialization, the developing countries, including China, still have considerable leeway in choosing their development paths. In this regard, the industrialized countries can play a very active role. They can act in a number of ways. For example, they can provide assistance in facilitating joint ventures. This can help their companies profit from China's booming economy and, at the same time, help the Chinese companies employ more advanced foreign technology and thereby use energy more efficiently.

In terms of historical CO_2 emissions, Table 3.16 shows that total CO_2 emissions from fossil fuels in China rose from 358.60 MtC in 1980 to 586.87 MtC in 1990, with an average annual growth rate of 5%. This means that China ranks second in global CO_2 emissions if the Soviet emissions are distributed over the new independent republics. But on a per capita basis, China's CO_2 emissions of 0.5 tC in 1990 were very low by comparison with the world average. With regard to the breakdown of CO_2 emissions by fuel, not surprisingly, coal predominates, accounting for 83.4% of the total emissions in 1990. Analysis of the contribution to CO_2 emissions growth over the period 1980-90 has shown that China's economic growth measured in per capita GDP was an overwhelming factor, alone resulting in an increase of 314.67 MtC. Population expansion was responsible for an increase of 68.27 MtC. The change in fossil fuel mix contributed to an increase of 5.33 MtC. By contrast, a reduction of 154.67 MtC was achieved because of great success in energy conservation. The penetration of carbon-free fuels also contributed to a small reduction in emissions.

Finally, it has been indicated that, driven by the threat of further degradation of the environment and the harmful economic effects of energy shortages, China is already determined to push energy conservation and enhanced energy efficiency in general and more efficient coal usage in particular. A number of policy measures, which have been and will continue to be implemented, have been outlined. These are the so-called 'no-regrets' measures in the sense that they are taken without considering the greenhouse effect. It has been shown that success in the implementation of these measures will depend largely on the extent to which a reform of China's energy pricing will be carried out. With respect to reducing CO_2 emissions, because the 'regrets' policies are often costly, 'getting prices right' and implementing the 'no-regrets' actions in China should have priority over the imposition of a carbon tax. Moreover, the growing environ-

---

[2] The joint implementation mechanism has been put forward for limiting global GHG emissions. Considerations with respect to China will be summarized in the next three sections.

mental concern built into both international and national programmes and China's rapid integration into the world economy tend to make China more amenable to international cooperation on the environment. All these, combined with the differentiated responsibility for abating carbon emissions between the industrialized and developing countries, suggest that the implementation of the 'no-regrets' measures will be accelerated if curbing global CO_2 emissions requires special action on China's part. With such efforts, China may keep its per capita CO_2 emissions well below the world average level without jeopardizing its economic development. This can be seen as a reasonable and achievable target for China.

9.2 Macroeconomic analysis of CO_2 emission limits for China

In Chapter 4, it has been argued that a computable general equilibrium (CGE) approach is generally considered an appropriate tool for analysing the economic impacts of limiting CO_2 emissions. Thus, a time-recursive dynamic CGE model of the Chinese economy has been designed, which has been described in Chapter 5 and been calibrated in Chapter 6. This CGE model operates by simulating the operation of markets for factors, products and foreign exchange, with equations specifying supply and demand behaviour across all markets. The model includes ten producing sectors and distinguishes four energy inputs. The CGE model is made up of the following nine blocks: production and factors, prices, income, expenditures, investment and capital accumulation, foreign trade, energy and environment, welfare, and market clearing conditions and macroeconomic balances. The model allows for endogenous substitution among energy inputs and alternative allocation of resources as well as endogenous determination of foreign trade and household consumption in the Chinese economy in order to cope with the carbon limits at both sectoral and macroeconomic levels. The model is also able to calculate the resulting welfare impacts. Furthermore, the CGE model incorporates an explicit tax system. This makes it suitable for estimating the 'double dividend' from the imposition of a carbon tax that is incorporated as a cost-effective means of limiting CO_2 emissions. Finally, the model is solved directly with a numerical solution technique included in GAMS.

Using this CGE model, a baseline scenario for the Chinese economy has first been developed under a set of assumptions about the exogenous variables. The calculations show that a rapid growth of the Chinese economy will take place until the year 2010. Consequently, this will lead to increased energy consumption and hence CO_2 emissions, despite substantial energy efficiency improvement. Moreover, a comparison with other studies for China has shown that of all the models considered, our estimates of the baseline CO_2 emissions represent the most plausible case from the point view of CO_2 emissions.

Then, using the time-recursive dynamic CGE model and assuming that carbon tax revenues are retained by the government, Section 7.3 analyses the implications of two scenarios under which China's CO_2 emissions in 2010 will be cut by 20% and 30% respectively relative to the baseline. The two emission targets are less restrictive in that they are not compared with the level of emissions in a single base year, but with the baseline CO_2 emissions in 2010, the latter being 2.46 times those in 1990. The main findings are as follows.

First, a larger absolute cut in CO_2 emissions will require a higher carbon tax. A higher tax also implies higher prices of fossil fuels. Moreover, carbon tax rises at an increasing rate as the target of CO_2 emissions becomes more stringent, indicating that large reductions in carbon emissions can only be achieved by ever-larger increases in carbon taxes and hence prices of fossil fuels.

Second, even under the two less restrictive carbon emission scenarios, China's GNP drops by 1.5% and 2.8% and its welfare measured in Hicksian equivalent variation drops by 1.1% and 1.8% respectively in 2010 relative to the baseline, indicating that the associated GNP and welfare losses tend to rise more sharply as the degree of the emission reduction increases. Given the often reported losses of 1-3 per cent of GDP in industrialized countries under very restrictive carbon limits, the results also support the general finding from global studies that China would be one of the regions hardest hit by carbon limits.

Third, although aggregate gross production tends to decrease at an increasing rate as the carbon dioxide emission target becomes more stringent, changes in gross production vary significantly among sectors in both absolute and relative terms. Of the ten sectors considered, we found that the coal sector is affected most severely in terms of output losses under the two CO_2 constraint scenarios. Consequently, this will lead to a considerable decline in the sector's employment. This, combined with the fact that about three-quarters of existing coal reserves are concentrated in the northern part, suggests that special attention should be paid to the sectoral and regional implications when designing a domestic carbon tax.

Fourth, of the four adjustment mechanisms considered, lower energy input coefficients contribute to the bulk of energy reduction and hence CO_2 emissions in 2010 under the two CO_2 constraint scenarios, followed by a change in the structure of economic activity. With respect to the contributions to carbon abatement in 2010, although in relative terms energy consumption in the coal sector and the corresponding CO_2 emissions in 2010 are reduced most under both scenarios, in absolute terms, the largest reductions occur in the heavy industry.

Imposing a carbon tax will raise government revenues. How these revenues are used will affect the overall economic burden of the tax. In Section 7.4, we have computed the efficiency improvement of four indirect tax offset scenarios relative to the two tax retention scenarios above. The four simulations labelled as Reforms 1a, 1b, 2a and 2b respectively show that if these revenues were used to offset reductions in indirect taxes, the negative impacts of carbon taxes on GNP and welfare would be reduced. Moreover, as shown by Reforms 2a and 2b as well as in Figures 7.2 and 7.3, the efficiency improvement tends to rise as the target of CO_2 emissions becomes more stringent (that is, fossil fuels are taxed more heavily by carbon taxes). This suggests that it would become more worthwhile to lower indirect taxes in order to reduce the adverse effects of a carbon tax.

In Section 7.5, a comparison with other studies for China has been made. It has been indicated that our estimates of the reduction in GNP growth are higher than those by GLOBAL 2100 and GREEN in order to achieve the same percentage of reductions in CO_2 emissions relative to the baseline. This difference might be related to three factors. First, our baseline of carbon emissions is higher than that in GLOBAL 2100 and GREEN. Second, our model is relatively disaggregated compared with both GLOBAL 2100 and GREEN. This implies less substitutability in our model, leading to higher economic costs. Third, model types matter. While in our single-country model one branch of industry is estimated to be negatively affected under the carbon constraints, this would not always be the case in a global model such as GREEN because of the relative improvement in Chinese branch goods' competitiveness via trade reallocation. The differing effects brought about by the imposition of unilateral carbon taxes or regional carbon taxes could be part of the explanation for the higher GNP losses in our model.

With respect to the carbon taxes required to achieve the same percentage of carbon reductions in 2010 relative to the baseline, our estimates are on the one hand higher than those by GREEN. This is because GREEN has a smaller size of the gap between the uncontrolled emissions and the emission target, and because GREEN has lower baseline prices of fossil fuels. On the other hand, our estimates are lower than those by GLOBAL 2100. This is because GLOBAL 2100 assumes lower values of the AEEI parameter, and because GLOBAL 2100 considers limited options for reducing CO_2 emissions and overstates the costs of some important alternative low carbon-polluting energy technologies. Moreover, comparing carbon tax levels across the regions considered shows that the carbon taxes required in China in order to achieve the same percentage of emission reductions relative to the baseline are much lower than those for both the industrialized countries and the world average. This provides the economic rationale for the industrialized countries to develop joint implementation (JI) projects with China. This will not only help China, which is becoming an important source of future CO_2 emissions, alleviate the suffering from possible future carbon limits, but also act to lower the costs of undertaking carbon abatement in the industrialized countries. Worldwide, this will achieve global carbon abatement at a lower overall cost than would otherwise have been the case.

Finally, it should be pointed out that although the arguments in this book are in favour of JI deals between the industrialized countries and developing countries (including China) on grounds of both cost-effectiveness and environmental protection, it is important to bear in mind that there would be more severe technical and institutional challenges posed by such deals than those deals between the industrialized countries. This is partly because there are today no national emission targets for the developing countries. Moreover, it is unlikely that these countries will adopt binding targets in the near future. This underlines the great uncertainty of obtaining an accurate evaluation of emissions for JI deals. Moreover, stringent monitoring and verification of emissions reductions of JI projects, and vigorous enforcement of contractual obligations will be essential if JI is to work. All this, combined with the lack of institutional framework and political instability in the developing countries, will lead to relatively high transaction costs of JI projects. All this will no doubt undermine the cost-effectiveness of JI deals.

9.3   Cost-effective analysis of carbon abatement options in China's electricity sector

Chapter 8 aims to shed light on technological aspects of carbon abatement in the electricity sector. Thus, it is devoted to satisfying electricity planning requirements in the CO_2 context. There were at least four reasons for studying the electricity sector. First, the electricity sector is a major consumer of primary energy. Second, there are many more interfuel substitution possibilities in the sector compared with other sectors. Third, the results obtained from studying the electricity sector can contribute to forming a necessary basis for China's development of JI projects with other countries. Fourth, time constraints did not allow studying the Chinese energy system as a whole.

For a cost-effective analysis of carbon abatement options in China's electricity sector, a technology-oriented optimization model for power system expansion planning has been developed. This model has been adapted from the MARKAL model. It chooses the minimization of discounted cost over the entire planning horizon as its objective function and incorporates a number of power-related constraints adopted by MARKAL. In the power planning model that has been programmed in GAMS, 15 types of power plants are represented in terms of their technical, economic and environmental parameters. Specifically, these parameters include initial investment, transmission and distribution costs, fixed and variable operating and maintenance costs, generation efficiency, plant utiliz-ation, plant lifetime, construction lead times, fuel requirements, dates of introduc-tion, maximum rates of expansion, and emissions coefficients. The model allows for substitution from high-carbon fossil fuels and technologies towards low-carbon and carbon-free counterparts and for interactions between periods to cope with carbon limits.

In order to examine the cost-effectiveness of the 15 types of power plants considered and hence to shed light on the priorities of abatement investments, a comparison of the power plants has been made in terms of both the levelized cost of generation and the marginal cost of CO_2 reduction at a 10% discount rate. Such a comparison shows that renewable energy plants, such as wind and PV plants, are still too costly in comparison with conventional coal and hydroelectric plants, although they can generate electricity without a directly parallel production of CO_2 emissions. It has also been found that large coal-fired plants (200 MW and above) and hydroelectric plants have negative marginal costs of CO_2 reduction in comparison with small coal-fired plants (<200 MW), the latter being chosen as the reference since they represent a worst-case option from the point of view of CO_2 emissions (see Tables 8.2 and 8.4). Thus, these plants should be given priority in carbon abatement investments. In practice, however, in order to promote the moving up to large units, efforts should be directed towards resolving a variety of problems, including the lack of domestic capacity for manufacturing large units, the difficulties in mobilizing the necessary large investment resources, and the disappointing performance of domestically-produced large units.

Using the power planning model and based on the macroeconomic results obtained through the CGE model concerning GNP growth and energy demand, a baseline scenario for electricity supply over the period to 2010 has been developed. The baseline scenario is characterized by a rapid growth of the power industry, with an average annual net add-up of 14 GW from 1990 to 2000 and 26 GW thereafter to 2010. Accordingly, capital investment in the electricity sector as a share of GNP is calculated to go up from the current level. With respect to a breakdown of total generating capacity by plant, the calculations show that coal-fired power plants still predominate, although the role of alternatives is growing. Accordingly, the amount of coal consumed for electricity generation grows rapidly, thus increasing its share in total coal consumption. This will lead to an increase in CO_2 emissions within the power sector itself, although the decreasing direct use of coal will alleviate the environmental impacts of coal use as a whole. With respect to the scale of coal plants, it has been shown that more large units are expected to be put into operation during the period under consideration compared with the current composition of plants. This will bring the average gross coal consumption of coal-fired plants down. Besides, the calculations show that although the share of nuclear power in total generating capacity is expected to rise to 0.76% in 2000 and to 3.45% in 2010, there is little prospect of dramatic increases during the period under consideration. It therefore follows that, in the short to medium term (before 2010), China has little alternative but to rely on coal for power generation, given the fact that long lead times and high capital costs pose difficulties for the expansion of both hydropower and nuclear power as well as renewable power to meet the projected rapidly-increasing electricity demand.

In Section 8.8, the impacts have been examined of a 65% rise in the price of coal as a result of the imposition of a carbon tax of 205 yuan per ton of carbon in order to achieve a 20% cut in CO_2 emissions in 2010. In this regard, three alternative policy simulations have been investigated, all of which are relative to the baseline. Simulation 1 shows the accelerated moving up to large coal-fired units, Simulation 2 the accelerated expansion of hydroelectric power, and Simulation 3 the accelerated expansion of nuclear power. Of the three simulations, the calculations for Simulation 1 suggest the largest reduction in average gross coal consumption of coal-fired plants, the smallest increase in capital investment, but the smallest reduction in CO_2 emissions. The calculations for Simulation 2 suggest the highest increase in capital investment, but the largest reduction in CO_2 emissions. The calculations for Simulation 3 suggest that capital investment and reduction in CO_2 emissions lie in between Simulations 1 and 2, but are closer to those for Simulation 2. With respect to the effects on the GNP losses and the price of electricity, the calculations through the CGE model show that there are slight differences among the three alternative policy simulations. This, combined with the effectiveness in a reduction in CO_2 emissions, would suggest the accelerated expansion of hydroelectric power and nuclear power in the carbon abatement investments (of course, the safety and radioactive waste disposal of nuclear power plants and the ecological damages from constructing large hydroelectric power plants must be taken into account). This finding is in line with the government investment priority regarding the electricity sector, which has been set with aims to reduce pressure on transportation and air pollution but without considering the greenhouse effect. It suggests that the development of hydroelectric power and nuclear power needs to be accelerated in China if curbing CO_2 emissions is taken into account. This implies that it is in China's interest to consider JI projects for increased generation efficiency and interfuel substitution in the electricity sector. This is viewed as a precondition for the development of JI projects with China. However, given the breadth of the subject of JI and its close linkage with national sovereignty, global political agenda, national development priorities, access to advanced technologies and development assistance, and the ongoing tension over the responsibilities of different parties to the FCCC, it is expected that a wide and successful implementation of these projects will be conditional upon the consensus on a variety of operational issues such as transaction costs, the form of JI, criteria for JI, the establishment of baselines against which the effects of JI projects can be measured, and the verification of emissions reductions of JI projects. At present, although some JI projects are under discussion, the Chinese government has not approved any JI projects, and JI remains virtually unknown to the majority of social and economic sectors in China as in most developing countries. It is, therefore, unrealistic to expect that JI projects with China work as smoothly and fast as the industrialized countries wish. This underlines the need to promote JI through pilot projects and capacity building in China in order to make JI gain ground and provide mutual benefits to all the parties involved.

9.4   Relevance and potential use of the results from this study for science and policy-making

This study provides the economic rationale for the industrialized countries to develop JI projects with China in order to lower their costs of undertaking carbon abatement and fulfil part of their emission abatement commitments. Moreover, the study discusses in some detail potential areas for JI projects with China. It has been shown that the most potential areas of interest to China are related to those activities and options aimed at: 1) improving the efficiency of energy use, particularly at energy-intensive energy sectors (for example, iron and steel industry, chemical industry, building materials industry, and power industry) and devices (for example, industrial boilers); 2) pushing efficient use of coal through increasing proportion of raw coal washed; popularizing domestic use of coal briquette; substitution of direct burning of coal by electricity through development of large-size, high-temperature and high-pressure efficient coal-fired power plants; expanding district heating systems and developing cogeneration; increased penetration of town gas into urban households; and through development of environmentally sound coal technologies; 3) speeding up the development of hydropower and nuclear power; and 4) developing renewables. These emission-abating options, though aimed at reducing GHG emissions, will contribute to the reductions of local pollutants and thus will be beneficial to a more sustainable development of the Chinese economy. It is the secondary benefits that may provide some incentives for China to cooperate with the industrialized countries on JI projects.

Until now, the Chinese government has ratified the United Nations Framework Convention on Climate Change and China's Agenda 21, the latter serving as a white paper of China's population, environment and development in the 21st century. China has also made great efforts to abolishing current subsidies for energy consumption, reducing barriers to trade and to protecting intellectual property rights in order to facilitate the transfer and spread of economically viable low-carbon or carbon-free advanced energy technologies. This at least indicates the Chinese government's genuine concern about the potential impacts of climate change on China and its willingness to take all possible measures to limit the growth of its own per capita GHG emissions. At present, China is beginning the process of developing an appropriate response strategy for climate change. Several projects have been initiated that deal with various aspects of climate change. Nevertheless, in China systematic and comprehensive research on the economic implications of carbon abatement for the Chinese economy is still in its infancy. In this regard, the results from this study, although still preliminary at this stage, should be of interest to energy and environmental policymakers in China and to those who seek to advise them. Moreover, the present study shows, among other things, not only the rationale for the JI mechanism, but also what kinds of JI projects are in China's interest. In the international arena, this is of particular interest to parties that are interested in obtaining China's cooperation in addressing regional and global environmental problems and in promoting sustainable development of the Chinese economy. For energy and environmental economists interested in quantifying economic impacts of energy and environ-mental policies and in carrying out a cost-effective analysis of options for limiting pollutant emissions, the application of the CGE modelling and the dynamic optimization approaches used in this study may be of interest. For the scientific community, given the fact that the extent of China's cooperation on JI will to some extent depend on the certainties about climate change, this study underlines the need to continue its efforts to clarify the scientific basis for climate change problem in order to lower the uncertainties about its magnitude, timing and regional patterns. Besides, the databases developed and the modelling exercise carried out within the framework of the present study provide a suitable basis for further study.

9.5   Suggestions for further work

The current CGE model provides a suitable and flexible basis for analysing the economic impacts of compliance with CO_2 emission limits. Nevertheless, there are some areas where there is a need for further methodological and empirical work in order to enrich the policy relevance of this study. This will also make the new version of the CGE model able to be used as a generalized tool for economic analysis of energy and environmental policy in China.

First, it would be desirable to incorporate an intertemporal optimization structure into the current CGE model. In the CO_2 context, there are at least two reasons for justifying such a change in the model structure. First, given that greenhouse gases are the so-called 'stock pollutants', the effectiveness of abatement policies should thus be judged in terms of a reduction in accumulated CO_2 emissions during the period under consideration, rather than in terms of the emissions rate at any particular point in time. Second, imposing a cumulative carbon limit provides an additional degree of policy flexibility, since it allows a country to optimize its time path of carbon emissions.

Second, it would be useful to incorporate estimates of the secondary benefits from reduced CO_2 emissions. Climate policy measures not only have a favourable climate effect, but also contribute to the reduction of local pollutants. Therefore, they help to solve local environmental problems, which are often regarded by the developing countries (including China) as their own environmental priorities. By including local pollutants, such as SO_2, NOx and particulates from fossil fuel burning, we can estimate the amount of reduced local pollutants from carbon abatement. If information about the relationship between the volume of local pollutants and the resulting environmental damages would become sufficiently available at some time in the future, this should be included in the current CGE model. So we can estimate the avoided environmental damages, that is, the secondary benefits from reduced local pollution that carbon control brings about. Given the fact that the Chinese government is more concerned with the avoidance of local negative externalities, the additional information about the secondary benefits may provide some incentives for China to take action to limit its future carbon emissions.

Third, it would be desirable to incorporate other greenhouse gases but CO_2. The current CGE model focuses only on CO_2 emissions because of their dominant contribution to global warming. But it should be emphasized that including not just CO_2 but also other greenhouse gases will induce more cost-effective options for controlling global warming.

Fourth, it would be desirable to incorporate aspects of income distribution if the required data become available. Although the current CGE model is appropriate for calculating the economic costs of a carbon tax, it cannot be used to analyse its distributional effects because of the lack of data on income distribution. However, distributional aspects are considered to be important when designing a domestic carbon tax.

Fifth, it would be desirable to incorporate a more realistic institutional structure. The Chinese economy is currently undertaking a historical transition from central planning to free market. The most distinctive feature of the transitional economy is the coexistence of diminishing central planning and rapidly expanding market. Including an explicit representation of the two-tier prices and plan/market flows will make our model have not only the more realistic representation of the Chinese economy but also the extended ability to analyse plan-related policy options such as adjustments in plan prices and output quantities. In view of possible future greenhouse gas emission limits, this modification will allow us to examine the economy-wide effects of eliminating energy market distortions arising from subsidies in order to use energy more efficiently and hence to lower CO_2 emissions.

Sixth, it would be desirable to extend the time horizon beyond 2010, given the long-term nature of the climate change issue. The longer the horizon, the more attention should be paid to the treatment of backstop technologies, which are crucial to limiting the carbon tax level required and thus reducing the economic costs incurred for compliance with emission limits.

References[1]

Adelman, I. and S. Robinson (1978), *Income Distribution Policy in Developing Countries: A Case Study of Korea*, Oxford University Press, Oxford.

Albouy, Y. (1991), *Coal Pricing in China: Issues and Reform Strategy*, World Bank Discussion Papers No. 138, The World Bank, Washington, DC.

Anderson, K. and R. Blackhurst (eds)(1992), *The Greening of World Trade Issues*, Harvester Wheatsheaf, New York.

Ayres, R.U. and J. Walter (1991), The Greenhouse Effect: Damages, Costs and Abatement, *Environmental and Resource Economics*, Vol. 1, pp. 237-70.

Bai, Xianhong (1991), Position and Strategy on Formulating the International Climate Change Framework Convention, in *Environmentally Sound Coal Technologies: Policy Issues and Options*, Report of the International Conference on Coal and Environment, Beijing, December 2-6, pp. 170-84.

Barker, T. (1992), *The Carbon Tax: Economic and Policy Issues*, Nota Di Lavoro 21.92, Fondazione Eni Enrico Mattei, Milano.

Barker, T. (1993), *Secondary Benefits of Greenhouse Gas Abatements: the Effects of a UK Carbon/Energy Tax on Air Pollution*, Nota Di Lavoro 32.93, Fondazione Eni Enrico Mattei, Milano.

Barker, T., Baylis, S. and P. Madsen (1993), A UK Carbon/Energy Tax: The Macroeconomic Effects, *Energy Policy*, Vol. 21, No. 3, pp. 296-308.

Barrett, S. (1990), The Problem of Global Environmental Protection, *Oxford Review of Economic Policy*, Vol. 6, No. 1, pp. 68-79.

Barrett, S. (1991), Global Warming: Economics of a Carbon Tax, in D.W. Pearce (ed), *Blueprint 2: Greening the World Economics*, Earthscan, London, pp. 30-52.

Barrett, S. (1995), *The Strategy of Joint Implementation in the Framework Convention on Climate Change*, United Nations Conference on Trade and Development, Geneva.

Bates, R.W. and E.A. Moore (1992), *Commercial Energy Efficiency and the Environment*, Policy Research Working Papers No. 972, The World Bank, Washington, DC.

Baumol, W.J. and W.E. Oates (1971), The Use of Standards and Prices for Protection of the Environment, *Swedish Journal of Economics*, Vol. 73, No. 1, pp. 42-54.

Baumol, W.J. and W.E. Oates (1988), *The Theory of Environmental Policy*, Cambridge University Press, Cambridge, 2nd Edition.

---

[1] Articles marked with * are written in Chinese.

Beauséjour, L., Gordon, L. and M. Smart (1995), A CGE Approach to Modelling Carbon Dioxide Emissions Control in Canada and the United States, *World Economy*, Vol. 18, No. 3, pp. 457-88.

Beaver, R. (1993), Structural Comparison of the Models in EMS 12, *Energy Policy*, Vol. 21, No. 3, pp. 238-48.

Bellman, R.E. (1957), *Dynamic Programming*, Princeton University Press, Princeton, New Jersey.

Berger, C., Dubois, R., Haurie, A., Lessard, E., Loulou, R. and J.-P. Waaub (1991), *Canadian MARKAL: An Advanced Linear Programming System for Energy and Environmental Modelling*, GERAD, Montreal, Canada.

Bergman, L. (1980), Energy Policy in a Small Open Economy: The Case of Sweden, RR-78-16, IIASA, Laxenberg, Austria.

Bergman, L. (1982), A System of Computable General Equilibrium Models for a Small Open Economy, *Mathematical Modelling*, Vol. 3, pp. 421-35.

Bergman, L. (1988), Energy Policy Modeling: A Survey of General Equilibrium Approaches, *Journal of Policy Modeling*, Vol. 10, No. 3, pp. 377-99.

Bergman, L. (1990), The Development of Computable General Equilibrium Modeling, in L. Bergman, D.W. Jorgenson and E. Zalai (eds), pp. 3-30.

Bergman, L. (1991), General Equilibrium Effects of Environmental Policy: A CGE-Modeling Approach, *Environmental and Resource Economics*, Vol. 1, No. 1, pp. 43-61.

Bergman, L., Jorgenson, D.W. and E. Zalai (eds)(1990), *General Equilibrium Modeling and Economic Policy Analysis*, Basil Blackwell, Oxford.

Bergman, L. and S. Lundgren (1990), General Equilibrium Approaches to Energy Policy Analysis in Sweden, in L. Bergman, D.W. Jorgenson and E. Zalai (eds), pp. 351-82.

Berndt, E.R. and D.O. Wood (1975), Technology, Prices and the Derived Demand for Energy, *Review of Economics and Statistics*, Vol. 57, No. 3, pp. 259-68.

Blitzer, C.R., Eckaus, R.S., Lahiri, S. and A. Meeraus (1992), *Growth and Welfare Losses from Carbon Emissions Restrictions: A General Equilibrium Analysis for Egypt*, Policy Research Working Papers No. 963, The World Bank, Washington, DC.

Boadway, R.W. and N. Bruce (1984), *Welfare Economics*, Basil Blackwell, Oxford.

Boero, G., Clarke, R. and L.A. Winters (1991), *The Macroeconomic Consequences of Controlling Greenhouse Gases: A Survey*, Environmental Economics Research Series, UK Department of the Environment, HMSO, London.

Bohm, P. (1993), Incomplete International Cooperation to Reduce CO_2 Emissions: Alternative Policies, *Journal of Environmental Economics and Management*, Vol. 24, No. 3, pp. 258-71.

Bohm, P. (1994), On the Feasibility of Joint Implementation of Carbon Emissions Reduction, in A. Amano *et al.* (eds), *Climate Change: Policy Instruments and their Implications, Proceedings of the Tsukuba Workshop of IPCC Working Group III*, Tsukuba, Japan, January 17-20, pp. 181-98.

Borges, A.M. (1986), Applied General Equilibrium Models: An Assessment of Their Usefulness for Policy Analysis, *OECD Economic Studies*, No. 7, pp. 7-43.

Bovenberg, A.L. (1985), The General Equilibrium Approach: Relevant for Public Policy?, in *The Relevance of Public Finance for Policy Making*, Proceeding of the 41st Congress of the International Institute of Public Finance, Madrid, pp. 33-43.

Bovenberg, A.L. (1994), *Environmental Policy, Distortionary Labor Taxation, and Employment: Pollution Taxes and the Double Dividend*, CentER for Economic Research, Tilburg University, Tilburg, The Netherlands.

Bovenberg, A.L. and R.A. de Mooij (1994), Environmental Taxation and Labor Market Distortions, *European Journal of Political Economics*, Vol. 10, pp. 655-83.

British Petroleum (1990), *BP Statistical Review of World Energy*, London.

British Petroleum (1993), *BP Statistical Review of World Energy*, London.

Brooke, A., Kendrick, D. and A. Meeraus (1988), *GAMS: A User's Guide*, The Scientific Press, San Francisco.

Brown, K. and D. Pearce (1994), The Economic Value of Non-Market Benefits of Tropical Forests: Carbon Storage, in J. Weiss (ed.), *The Economics of Project Appraisal and the Environment*, Edward Elgar, Aldershot, England, pp. 102-23.

Bruggink, J.J.C. (1995), Assessment Report on NRP Subtheme 'International Instruments for Climate Change Policy', in S. Zwerver, R.S.A.R. van Rompaey, M.T.J. Kok and M.M. Berk (eds), *Climate Change Research: Evaluation and Policy Implications*, Elsevier Science Publishers, Amsterdam, pp. 1293-313.

Burniaux, J.M., Martin, J.P., Nicoletti, G. and J.O. Martins (1991), *The Costs of Policies to Reduce Global Emissions of CO_2: Initial Simulation Results with GREEN*, Working Papers No. 103, Department of Economics and Statistics, OECD, Paris.

Burniaux, J.M., Martin, J.P., Nicoletti, G. and J.O. Martins (1992), *GREEN - A Multi-Sector, Multi-Region General Equilibrium Model for Quantifying the Costs of Curbing CO_2 Emissions: A Technical Manual*, Working Papers No. 116, Department of Economics and Statistics, OECD, Paris.

Byrd, W.A. (1989), Plan and Market in the Chinese Economy: A Simple General Equilibrium Model, *Journal of Comparative Economics*, Vol. 13, No. 2, pp. 1-77-204.

Capros, P., Karadeloglou, P. and G. Mentzas (1990a), An Empirical Assessment of Macroeconometric and CGE Approaches in Policy Modeling, *Journal of Policy Modeling*, Vol. 12, No. 3, pp. 557-85.

Capros, P., Karadeloglou, P. and G. Mentzas (1990b), The MIDAS Energy Model and the MIDAS-HERMES Linked System: Methodology and Application to Carbon-Tax System, Paper Presented at the Workshop on Economic/Energy/Environmental Modeling for Climate Policy Analysis, Washington, DC, October 22-23.

Carraro, C., Lanza, A. and A. Tudini (1994), *Technological Change, Technology Transfers and the Negotiation of International Environmental Agreements*, Nota Di Lavoro 37.94, Fondazione Eni Enrico Mattei, Milano.

CEC (1991), *A Community Strategy to Limit Carbon Dioxide emissions and to Improve Energy Efficiency*, Commission of the European Communities (CEC), Brussels.

*CERS (1986), *The Present Situation and Prospect for China's Energy*, China Energy Research Society (CERS), Beijing.

*CERS (ed)(1992), *Market Economy and China Energy Development Strategy*, China Energy Research Society (CERS), Atomic Energy Press, Beijing.

Chambers, R.G. (1988), *Applied Production Analysis: A Dual Approach*, Cambridge University Press, Cambridge.

Chandler, W.U., Dadi, Zhou and J. Hamburger (1993), US-China Cooperation for Global Environmental Protection, *International Journal of Global Energy Issues*, Vol. 5, Nos. 2/3/4, pp. 169-76.

Chen, Kang (1990), *An Economic Model of China*, PhD Dissertation, University of Maryland.

Chen, Yingrong (1991), Renewables in China: Case Study, *Energy Policy*, Vol. 19, No. 11, pp. 892-96.

Chen, Zongming et al. (1993), *Future Supply of Petroleum and Natural Gas in China*, China Petroleum and National Gas Corporation, Beijing.

Climate Network Europe (ed)(1994), *Joint Implementation from a European NGO Perspective*, Brussels.

Cline, W.R. (1991), Scientific Basis for the Greenhouse Effect, *The Economic Journal*, Vol. 101, pp. 904-19.

Cline, W.R. (1992), *The Economics of Global Warming*, Institute of International Economics, Washington, DC.

Cline, W.R. (1994), *Costs and Benefits of Greenhouse Abatement: A Guide to Policy Analysis*, in OECD (1994), pp. 87-105.

Codoni, R., Park, H.-C. and K.V. Ramani (1985), *Integrated Energy Planning: A Manual, Vol II Energy Supply*, Asian and Pacific Development Centre, Kuala Lumpur.

Conrad, K. and M. Schröder (1991), The Control of CO_2 Emissions and its Economic Impact: An AGE Model for a German State, *Environmental and Resource Economics*, Vol. 1, No. 3, pp. 289-312.

Conrad, K. and M. Schröder (1993), Choosing Environmental Policy Instruments Using General Equilibrium Models, *Journal of Policy Modeling*, Vol. 15, Nos. 5&6, pp. 521-43.

Customs General Administration (1985), *The Official Customs Guide 1985/86*, Longman, London.

*Customs General Administration (1988), *Summary Survey of China's Customs Statistics 1987*, Knowledge Press, Beijing.

Dasgupta, P., Kriström, B. and K.-G. Mäler (1995), Current Issues in Resource Accounting, in P.-O. Johansson, B. Kriström and K.-G. Mäler (eds), *Current Issues in Environmental Economics*, Manchester University Press, Manchester.

Dean, A. and P. Hoeller (1992), Costs of Reducing CO_2 Emissions: Evidence from Six Global Models, *OECD Economic Studies* No. 19, pp. 15-47.

Dervis, K., de Melo, J. and S. Robinson (1982), *General Equilibrium Models for Development Policy*, A World Bank Research Publication, The World Bank, Washington, DC.

Development Research Center (1993), *A Proto-Type CGE Model for the Chinese Economy*, The State Council, Cooperated with Institute of Systems Science, Chinese Academy of Sciences, Beijing.

Dewatripont, M. and G. Michel (1987), On Closure Rules, Homogeneity and Dynamics in Applied General Equilibrium Models, *Journal of Development Economics*, Vol. 26, pp. 65-76.

Dixon, P.B. and B.R. Parmenter (1994), *Computable General Equilibrium Modelling*, Centre of Policy Studies, Monash University, Australia. Forthcoming in: Handbook in Computational Economics, North-Holland, Amsterdam.

Dixon, P.B., Parmenter, B.R., Sutton, J. and D.P. Vincent (1982), *ORANI: A Multisectoral Model of the Australian Economy*, North-Holland, Amsterdam.

Dornbusch, R. and J. Poterba (eds)(1991), *Global Warming: Economic Policy Responses*, MIT Press, Cambridge.

DRI (1991), *The Economic Impact of a Package of EC Measures to Control CO_2 Emissions*, Final Report Prepared for the CEC.

Dudek, D.J. and J.B. Wiener (1996), *Joint Implementation and Transaction costs under the Climate Change Convention*, Final Report Prepared for the Organization for Economic Cooperation and Development.

Edmonds, J. and J. Darmstadter (1990), Human Development and Carbon Dioxide Emissions: Part I Background, Reference Projection and Uncertainty Analysis, *International Journal of Global Energy Issues*, Vol. 2, No. 1, pp. 3-13.

Edmonds, J. and J. Reilly (1983), A Long-Term Global Energy-Economic Model of Carbon Dioxide Release from Fossil Fuel Use, *Energy Economics*, Vol. 4, pp. 74-88.

Ehrlich, P. and A. Ehrlich (1990), *The Population Explosion*, Touchstone, New York.

Enders, A. and A. Porges (1992), Successful Conventions and Conventional Success: Saving the Ozone Layer, in K. Anderson and R. Blackhurst (eds), pp. 130-44.

*Energy Research Institute (1991), CO_2 Emissions from Fossil Fuel Combustion and Reduction Countermeasures in China, *Bulletin of Energy Policy Research*, No. 2, pp. 4-7, State Planning Commission, Beijing.

Environmental Defense Fund (1993), *Joint Implementation: Sustainable Development through Trade in Environmental Commodities*, New York.

Ettinger, J. van, Jansen, T. and C. Jepma (1991), Climate, Environment and Development, *European Journal of Development Research*, Vol. 3, No. 1, pp. 108-32.

EWC/ANL/TU (1994), *National Response Strategy for Global Climate Change: People's Republic of China*, East-West Center (EWC), Argonne National Laboratory (ANL) and Tsinghua University (TU), Final Report to the Asian Development Bank.

Ezaki, M. and S. Ito (1993), Planning and the Free Market: A CGE Analysis of the Chinese Economy, presented at the International Conference on Macro-Economic modelling of China, Beijing, February 18-20.

Fankhauser, S. (1994), Valuing Climate Change: An Economic Assessment of Global Warming Impacts, presented at the NATO Advanced Workshop on the Economics of Atmospheric Pollution, Wageningen, The Netherlands, November 16-18.

Fankhauser, S. and D. McCoy (1995), Modelling the Economic Consequences of Environmental Policies, in H. Folmer, L. Gabel and J. Opschoor (eds), *Principles of Environmental and Resource Economics: A Guide for Students and Decision-Makers*, Edward Elgar, Aldershot, England, pp. 253-75.

Faucheux, S. and J.F. Noel (1991), Energy Analysis as a Means to Improve Environmental Taxation, Presented at the Annual Meeting of the EAERE, Stockholm, June 11-14.

Felder, S. and T.F. Rutherford (1993), Unilateral CO_2 Reductions and Carbon Leakage: The Consequences of International Trade in Oil and Basic Materials, *Journal of Environmental Economics and Management*, Vol. 25, pp. 162-76.

Fishbone, L.G. *et al.* (1983), *User's Guide for MARKAL (BNL/KFA Version 2.0): A Multi-Period, Linear-Programming Model for Energy Systems Analysis*, BNL 51701, Brookhaven National Laboratory and Associated Universities, Inc., Upton, New York.

Folmer, H., van Mouche, P. and S. Ragland (1993), International Environmental Problems and Interconnected Games, *Environmental and Resource Economics*, Vol. 3, pp. 313-35.

Fragnière, E. and A. Haurie (1995), *MARKAL-Geneva: A Model to Assess Energy-Environment Choices for a Swiss Canton*, Nota Di Lavoro 8.95, Fondazione Eni Enrico Mattei, Milano.

Fritsche, U. (1994), The Problems of Monitoring and Verification of Joint Implementation, in Climate Network Europe (ed), pp. 13-24.

Garbaccio, R.F. (1994), Reform and Structural Change in the Chinese Economy: A CGE Analysis, Presented at the American Economic Association - Chinese Economists Society Joint Session 'China in the World Economy (I)', Boston, January 3-5.

Gaskins, D. and J.P. Weyant (eds)(1995), *Reducing Global Carbon Emissions: Cost and Policy Options*, Stanford University Press, Stanford, California.

Ginsburgh, V.A. and J.L. Wealbroeck (1981), *Activity Analysis and General Equilibrium Modeling*, North-Holland, Amsterdam.

Ginsburgh, V.A. and J.L. Wealbroeck (1984), Planning Models and General Equilibrium Analysis, in H. Scarf and J. Shoven (eds), pp. 415-39.

Glomsrød, S., Vennemo, H. and T. Johnsen (1992), Stabilization of Emissions of CO_2: A Computable General Equilibrium Assessment, *Scandinavian Journal of Economics*, Vol. 94, No. 1, pp. 53-69.

Goldstein, G.A. (1991), *PC-MARKAL and the MARKAL Users Support System (MUSS)*, BNL-46319, Brookhaven National Laboratory and Associated Universities, Inc., Upton, New York.

Goulder, L.H. (1994), *Environmental Taxation and the 'Double Dividend': A Reader's Guide*, Department of Economics, Stanford University.

Goulder, L.H. (1995), Effects of Carbon Taxes in an Economy with Prior Tax Distortions: An Intertemporal General Equilibrium Analysis, *Journal of Environmental Economics and Management*, Vol. 29, pp. 271-97.

Grubb, M. (1989), *The Greenhouse Effect: Negotiating Targets*, The Royal Institute of International Affairs, London.

Grubb, M. (1993), The Costs of Climate Change: Critical Elements, in Y. Kaya, N. Nakicenovic, W.D. Nordhaus and F.L. Toth (eds), *Costs, Impacts and Benefits of CO_2 Mitigation*, IIASA CP-93-2, Austria, pp. 153-66.

Grubb, M., Edmonds, J., ten Brink, P. and M. Morrison (1993), The Costs of Limiting Fossil-Fuel CO_2 Emissions: A Survey and Analysis, *Annual Review of Energy and Environment*, Vol. 18, pp. 397-478.

Grubb, M., Koch, M., Thompson, K., Munson, A. and F. Sullivan (1993), *The 'Earth Summit' Agreements: A Guide and Assessment*, The Royal Institute of International Affairs/Earthscan, London.

Gunning, J.W. and M.A. Keyzer (1995), *Applied General Equilibrium Models for Policy Analysis*, in T.N. Srinivasan and J. Behrman (eds), *Handbook of Development Economics*, Vol. 3, pp. 2025-107, North-Holland, Amsterdam.

*Guo, Xiaomin *et al.* (1990), The Calculation of Economic Losses from Environmental Pollution in China, *China Environmental Science*, Vol. 10, No. 1, pp. 51-9.

Hahn, R.W. (1984), Market Power and Transferable Property Rights, *Quarterly Journal of Economics*, Vol. 99, pp. 753-65.

Haigh, N. (1989), New Tools for European Air Pollution Control, *International Environmental Affairs*, Vol. 1, pp. 26-37.

Hare, B. and A. Stevens (1995), Joint Implementation: A Critical Approach, in C.J. Jepma (ed), pp. 79-85.

Harris, R. (1984), Applied General Equilibrium Analysis of Small Open Economics with Scale Economies and Imperfect Competition, *American Economic Review*, Vol. 74, pp. 1016-32.

Haugland, T., Olsen, Ø. and K. Roland (1992), Stabilizing CO_2 Emissions: are Carbon Taxes a Viable Option? *Energy Policy*, Vol. 20, No. 5, pp. 405-15.

*He, Jiankun and Yingyun Lu (1992), China's Energy Demand and Role of Nuclear Energy in Future, in CERS (ed, 1992), pp. 92-101.

Hoel, M. (1991), Efficient International Agreements for Reducing Emissions of CO_2, *The Energy Journal*, Vol. 12, No. 2, pp. 93-108.

Hoel, M. (1992), International Environment Conventions: The Case of Uniform Reductions of Emissions, *Environmental and Resource Economics*, Vol. 2, pp. 141-59.

Hoeller, P. and J. Coppel (1992), Carbon Taxes and Current Energy Policies in OECD Countries, *OECD Economic Studies*, No. 19, pp. 167-93.

Hoeller, P., Dean, A. and J. Nicolaisen (1991), Macroeconomic Implications of Reducing Greenhouse Gas Emissions: A Survey of Empirical Studies, *OECD Economic Studies*, No. 16, pp. 45-78.

Hogan, W.W. (1990), Comments on Manne and Richels: 'CO_2 Emission Limits: An Economic Cost Analysis for the USA', *The Energy Journal*, Vol. 11, No. 2, pp. 75-84.

Hogan, W.W. and D.W. Jorgenson (1991), Productivity Trends and the Cost of Reducing CO_2 Emissions, *The Energy Journal*, Vol. 12, No. 1, pp. 67-86.

Houghton, J.T., Jenkins, G.J. and J.J. Ephraums (eds)(1990), *The IPCC Scientific Assessment*, Cambridge University Press, Cambridge.

Howe, H. (1975), Development of the Extended Linear Expenditure System from Simple Saving Assumptions, *European Economic Review*, Vol. 6, pp. 305-10.

*Hu, Zhaoyi (1991), *Study on Decision Support Indicator System of Electric Power Industry Development*, Electric Power Research Institute, Ministry of Energy, Beijing.

*Hu, Zhaoyi *et al.* (1990), *The Characteristics of China's Energy Supply and Demand and its Countermeasures of Energy Conservation*, Electric Power Research Institute, Ministry of Energy, Beijing.

*Hu, Zhaoyi *et al.* (1992), *Forecast for Flow of China's Electricity Generation in 2020*, Electric Power Research Institute, Ministry of Energy, Beijing.

Hussain, A. (1995), Discussion on Clarke and Winters: 'Energy Pricing for Sustainable Development in China', in I. Goldin and L.A. Winters (eds), *The Economics of Sustainable Development*, Cambridge University Press, Cambridge, pp. 230-235.

IEA (1989), *Energy Policies and Programmes of IEA Countries*, International Energy Agency (IEA), OECD, Paris.

IEA (1992), *Climate Change Policy Initiatives*, International Energy Agency (IEA), OECD, Paris.

Ingham, A. and A. Ulph (1991), Carbon Taxes and the UK Manufacturing Sector, in F. Dietz, F. van der Ploeg and J. van der Straaten (eds), *Environmental Policy and the Economy*, Elsevier, Amsterdam, pp. 197-239.

IPCC (1990), *Policymakers Summary of the Scientific Assessment of Climate Change*, WMO and UNEP.

IPCC (1991), *Climate Change: The IPCC Response Strategies*, WMO and UNEP.

Italianer, A. (1986), *The HERMES Model: Complete Specification and First Estimation Results*, Commission of the European Communities, Brussels.

Jackson, T. (1991), Least-Cost Greenhouse Planning: Supply Curves for Global Warming Abatement, *Energy Policy*, Vol. 19, No. 1, pp. 35-46.

James, D.E., Musgrove, A.R. del and K.J. Stocks (1986), Integration of an Economic Input-Output Model and a Linear Programming Technological Model for Energy System Analysis, *Energy Economics*, Vol. 7, No. 2, pp. 99-112.

Jepma, C.J. (1994), *Inter-nation Policy Co-ordination and Untying of Aid*, Ashgate Publishing Limited, Aldershot, England.

Jepma, C.J. (ed)(1995), *The Feasibility of Joint Implementation*, Kluwer Academic Publishers, Dordrecht, The Netherlands.

Johansen, L. (1960), *A Multi-sectoral Study of Economic Growth*, North-Holland, Amsterdam.

Jones, C.T. (1993), Another Look at US Passenger Vehicle Use and the Rebound-Effect from Improved Fuel Efficiency, *The Energy Journal*, Vol. 14, No. 4, pp. 99-110.

Jones, T. (1994), Operational Criteria for Joint Implementation, in OECD (1994), pp. 109-25.

Jorgenson, D.W. and P.J. Wilcoxen (1989), *Environmental Regulation and U.S. Economic Growth*, Discussion Paper Number 1458, Harvard Institute of Economic Research, Harvard University, Cambridge, Massachusetts.

Jorgenson, D.W. and P.J. Wilcoxen (1993a), Reducing US Carbon Emissions: An Econometric General Equilibrium Assessment, *Resources and Energy Economics*, Vol. 15, No. 1, pp. 7-25.

Jorgenson, D.W. and P.J. Wilcoxen (1993b), Reducing U.S. Carbon Dioxide Emissions: An Assessment of Different Instruments, *Journal of Policy Modeling*, Vol. 15, Nos. 5&6, pp. 491-520.

Kågeson, P. (1991), *Economic Instruments for Reducing Western European Carbon Dioxide Emission*, Swedish Environmental Advisory Council, Ministry of the Environment, Stockholm.

Kane, S., Reilly, J. and J. Tobey (1992), An Empirical Study of the Economic Effects of Climate Change on World Agriculture, *Climatic Change*, Vol. 21, pp. 17-35.

Karadeloglou, P. (1992), Energy Tax versus Carbon Tax: A Quantitative Macroeconomic Analysis with the HERMES/MIDAS Models, in F. Laroui and J.W. Velthuijsen (eds), pp. 127-52.

Kats, G. (1990), Slowing Global Warming and Sustainable Development: The Promise of Energy Efficiency, *Energy Policy*, Vol. 18, No. 1, pp. 25-33.

Kats, G. (1992), The Earth Summit: Opportunity for Energy Reform, *Energy Policy*, Vol. 20, No. 6, pp. 546-58.

Keepin, B. and G. Kats (1988), Greenhouse Warming: Comparative Analysis of Nuclear and Efficiency Abatement Strategies, *Energy Policy*, Vol. 16, No. 6, pp. 538-61.

Kleemann, M. and D. Wilde (1990), Intertemporal Capacity Expansion Models, *Energy*, Vol. 15, Nos. 7/8, pp. 549-60.

Kram, T. (1993a), *National Energy Options for Reducing CO_2 Emissions, Vol. I: The International Connection*, A Report of the IEA-ETSAP/Annex IV (1990-1993), ECN-C-93-101, Netherlands Energy Research Foundation (ECN), Petten, The Netherlands.

Kram, T. (ed)(1993b), *Proceedings of Seminar on 'Linking Technical Energy System Models with Macro-economic Approaches'*, Oxford, June 7-9.

Kram, T. (ed)(1994), *Proceedings of the 3rd Workshop of the IEA-ETSAP/Annex V*, ECN, Petten, the Netherlands, April 13-15.

Kuik, O., Peters, P. and N. Schrijver (eds)(1994), *Joint Implementation to Curb Climate Change: Legal and Economic Aspects*, Kluwer Academic Publishers, Dordrecht, The Netherlands.

Kverndokk, S. (1993), Global CO_2 Agreement: A Cost-Effective Approach, *The Energy Journal*, Vol. 14, No. 2, pp. 91-112.

Kypreos, S. (1990), *The Model MARKAL: Theory and Applications*, Paul Scherrer Institute, Villigen, Switzerland.

Larsen, B. and A. Shah (1992), *World Fossil Fuel Subsidies and Global Carbon Emissions*, Policy Research Working Papers No. 1002, The World Bank, Washington, DC.

Larsen, B. and A. Shah (1994), Global Tradeable Carbon Permits, Participation Incentives, and Transfers, *Oxford Economic Papers*, Vol. 46, Special Issue on Environmental Economics, pp. 841-56.

Laroui, F. and J.W. Velthuijsen (eds)(1992), *An Energy Tax in Europe*, SEO Report No. 281, Amsterdam.

Lee, D.R. and Misiolek, W.S. (1986), Substituting Pollution Taxation for General Taxation: Some Implications for Efficiency in Pollution Taxation, *Journal of Environmental Economics and Management*, Vol. 13, pp. 338-47.

Lee, H., Martins, J.O. and D. van der Mensburgghe (1994), Introduction to the GREEN Model, in J. Fr. Hake *et al.* (eds), *Advances in System Analysis: Modelling Energy-Related Emissions on a National and Global Level*, KFA, Jülich, Germany, pp. 263-98.

Leontief, W. (1970), Environmental Repercussions and the Economic Structure: An Input-Output Approach, *Review of Economics and Statistics*, Vol. 52, pp. 262-77.

*Li, Junsheng (1992), The Technology Transformation for Energy Conservation and Its Policy in China, in CERS (ed, 1992), pp. 147-57.

Li, X.Z., Yang, S.M. and J.H. He (1985), *The Structure of China's Domestic Consumption: Analyses and Preliminary Forecasts*, World Bank Staff Working Paper No. 755, The World Bank, Washington, DC 20433.

*Li, Z., Liu, Y.Q. and H.M. Liu (1990), The Status and Prospects of Nuclear Power in China's Energy System, in X.Q. Zhou and C.Z. Zhu (eds), *Forecast for Energy Demand in China to the year 2050*, Ministry of Energy, Beijing, pp. 160-95.

Lipietz, A. (1995), Enclosing the Global Commons: Global Environmental Negotiations in a North-South Conflictual Approach, in V. Bhaskar and A. Glyn (eds), *The North, The South and The Environment: Ecological Constraint and the Global Economy*, United Nations University Press, Tokyo; Earthscan, London, pp. 118-42.

*Liu, Jian (1989), General Talk on Energy Conservation, *Energy of China*, No.3, pp. 39-42.

Loske, R. and S. Oberthür (1994), Joint Implementation under the Climate Change Convention, *International Environmental Affairs*, Vol. 6, No. 1, pp. 45-58.

Lu, Yingzhong (1986), *Long-Term Energy Forecasting of PRC and Strategies in the Development of Nuclear Energy*, Working Paper at IEA/ORAU, Oak Ridge, Tennessee.

Lucas, R. (1976), Econometric Policy Evaluation: A Critique, *Journal of Monetary Economics*, Vol. 1, Supplement, pp. 19-46.

MacKerron, G. (1992), Nuclear Costs: Why Do They Keep Rising?, *Energy Policy*, Vol. 20, No. 7, pp. 641-52.

Mankiw, N.G. (1990), A Quick Refresher Course in Macroeconomics, *Journal of Economic Literature*, Vol. 28, pp. 1645-60.

Manne, A.S. (1992), *Global 2100: Alternative Scenarios for Reducing Carbon Emissions*, Working Papers No. 111, Department of Economics and Statistics, OECD, Paris.

Manne, A.S. (1994), International Trade: The Impact of Unilateral Carbon Emission Limits, in OECD (1994), pp. 193-205.

Manne, A.S. and R.G. Richels (1990), CO_2 Emission Limits: An Economic Cost Analysis for the USA, *The Energy Journal*, Vol. 11, No. 2, pp. 51-74.

Manne, A.S. and R.G. Richels (1991a), Global CO_2 Emission Reductions - the Impacts of Rising Energy Costs, *The Energy Journal*, Vol. 12, No. 1, pp. 87-107.

Manne, A.S. and R.G. Richels (1991b), International Trade in Carbon Emission Rights: A Decomposition Procedure, *American Economic Review*, Vol. 81, No. 2, pp. 135-9.

Manne, A.S. and R.G. Richels (1992), *Buying Greenhouse Insurance: The Economic Costs of CO_2 Emission Limits*, The MIT Press, Cambridge, Massachusetts.

Manne, A.S. and R.G. Richels (1993), The EC Proposal for Combining Carbon and Energy Taxes: The Implications for Future CO_2 Emissions, *Energy Policy*, Vol. 21, No. 1, pp. 5-12.

Manne, A. and R. Richels (1995), The Greenhouse Debate: Economic Efficiency, Burden Sharing and Hedging Strategies, *The Energy Journal*, Vol. 16, No. 4, pp. 1-37.

Manne, A.S. and C.-O. Wene (1992), *MARKAL-MACRO: A Linked Model for Energy-Economy Analysis*, BNL-47161, Brookhaven National Laboratory (BNL) and Associated Universities, Inc., Upton, New York.

Mansur, A. and J. Whalley (1984), Numerical Specification of Applied General Equilibrium Models: Estimation, Calibration and Data, in H. Scarf and J. Shoven (eds), pp. 69-127.

Martin, J.P., Burniaux, J.M., Nicoletti, G. and J.O. Martins (1992), The Costs of International Agreements to Reduce CO_2 Emissions: Evidence from GREEN, *OECD Economic Studies*, No. 19, pp. 93-121.

Martin, W. (1990), *Modelling the Post-Reform Chinese Economy*, China Working Paper No. 90/1, National Centre for Development Studies, Research School of Pacific Studies, The Australian National University, Canberra.

Martins, J.O., Burniaux, J.M. and J.P. Martin (1992), Trade and the Effectiveness of Unilateral CO_2-Abatement Policies: Evidence from GREEN, *OECD Economic Studies*, No. 19, pp. 123-40.

Martins, J.O., Burniaux, J.M., Martin, J.P. and G. Nicoletti (1993), The Costs of Reducing CO_2 Emissions: A Comparison of Carbon Tax Curves with GREEN, in OECD (1993a), pp. 67-94.

McKitrick, Ross R. (1995), *Welfare Effects of Alternative Revenue-Neutrality Rules for Canadian Carbon Taxes: An Econometric General Equilibrium Analysis*, Department of Economics, University of British Columbia, Vancouver.

Melo, J. de (1988), Computable General Equilibrium Models for Trade Policy in Developing Countries: A Survey, *Journal of Policy Modeling*, Vol. 10, No. 4, pp. 469-503.

Melo, J. de and S. Robinson (1989), Product Differentiation and the Treatment of Foreign Trade in Computable General Equilibrium Models of Small Economies, *Journal of International Economics*, Vol. 27, pp. 47-67.

Merrill, O.H. (1972), *Applications and Extensions of an Algorithm that Computes Fixed Points of Certain Upper Semi-continuous Point to Set Mapping*, PhD Dissertation, University of Michigan.

Metz, B. (1995), Joint Implementation: What the Parties to the Climate Convention Should Do about it, in C.J. Jepma (ed), pp. 163-75.

Merkus, H. (1992), *The Framework Convention on Climate Change: Some Thoughts on Joint Implementation*, CCD/Paper 11, Climate Change Division, Ministry of Housing, Spatial Planning and the Environment, The Netherlands.

Michaelowa, A. (1995), *Joint Implementation of Greenhouse Gas Reductions under Consideration of Fiscal and Regulatory Incentives*, HWWA-Report No. 153, Hamburg Institute for Economic Research (HWWA), Hamburg.

Miller, R.E. and P.D. Blair (1985), *Input-Output Analysis: Foundations and Extensions*, Prentice Hall, Englewood Cliffs.

*Ministry of Energy (1991a), *A Compilation of Electric Power Industry Statistics*, Beijing.

Ministry of Energy (1991b), *Electric Power Industry in China*, Beijing.

*Ministry of Energy (1992a), *A Compilation of Electric Power Industry Statistics*, Beijing.

Ministry of Energy (1992b), *Electric Power Industry in China*, Beijing.

Misiolek, W.S. and H.W. Elder (1989), Exclusionary Manipulation of Markets for Pollution Rights, *Journal of Environmental Economics and Management*, Vol. 16, pp. 156-66.

Monopolies and Mergers Commission (1981), *Central Electricity Generating Board: A Report on the Operation by the Board of its System for the Generation and Supply of Electricity in Bulk*, Presented to Parliament in Pursuance of

Section 17 of the Competition Act 1980, Her Majesty's Stationery Office, London.

Morris, S.C., Solomon, B.D., Hill, D., Lee, J. and G. Goldstein (1991), A Least Cost Energy Analysis of U.S. CO_2 Reduction Options, in J. Tester *et al.* (eds), *Energy and Environment in the 21st Century*, The MIT Press, Cambridge, MA, pp. 865-876.

Neary, J.P. and K.W.S. Roberts (1980), The Theory of Household Behaviour under Rationing, *European Economic Review*, Vol. 13, pp. 25-42.

Nordhaus, W.D. (1979), *The Efficient Use of Energy Resources*, Yale University Press, New Haven.

Nordhaus, W.D. (1990), Greenhouse Economics, *The Economist* (July 7), pp. 19-22.

Nordhaus, W.D. (1991a), The Cost of Slowing Climate Change: A Survey, *The Energy Journal*, Vol. 12, No. 1, pp. 37-65.

Nordhaus, W.D. (1991b), To Slow or not to Slow: the Economics of the Greenhouse Effect, *The Economic Journal*, Vol. 101, pp. 920-37.

Nordic Council of Ministers (1995), *Joint Implementation as a Measure to Curb Climate Change: Nordic Perspectives and Priorities*, TemaNord 534, Copenhagen.

OECD (1985), *Electricity in the IEA Countries: Issues and Outlooks*, International Energy Agency (IEA), Paris.

OECD (1989), *Projected Costs of Generating Electricity from Power Stations for Commissioning in the Period 1995-2000*, Paris.

OECD (1991), *Estimations of GHG Emissions and Sinks*, Final Report OECD Expert Meeting, Paris.

OECD (1992), *Proceedings of Workshop on Emissions Trading*, Paris.

OECD (1993a), *The Costs of Cutting Carbon Emissions: Results from Global Models*, Paris.

OECD (1993b), *The Use of Individual Quotas in Fisheries Management*, Paris.

OECD (1994), *The Economics of Climate Change, Proceedings of an OECD/IEA Conference*, Paris.

Ogawa, Y. (1991), Economic Activity and the Greenhouse Effect, *The Energy Journal*, Vol. 12, No. 1, pp. 23-35.

Owen, A.D. and P.N. Neal (1989), China's Potential as an Energy Exporter, *Energy Policy*, Vol. 17, No. 10, pp. 485-500.

Parikh, J.K. (1995), Joint Implementation and North-South Cooperation for Climate Change, *International Environmental Affairs*, Vol. 7, No. 1, pp. 22-41.

Pearce, D. (1990), Economics and the Global Environmental Challenge, *Journal of International Studies*, Vol. 19, No. 3, pp. 365-87.

Pearce, D. (1991), The Role of Carbon Taxes in Adjusting to Global Warming, *The Economic Journal*, Vol. 101, pp. 938-48.

Pearce, D. (1995a), *Blueprint 4: Capturing Global Environmental Value*, Earthscan, London.

Pearce, D. (1995b), Joint Implementation: A General Overview, in C.J. Jepma (ed), pp. 15-31.

Pearce, D. (1996), Global Environmental Value and the Tropical Forests: Demonstration and Capture, in W.L. Adamowicz and P. Boxall *et al.* (eds), *Forestry, Economics and the Environment*, Cab International, Wallingford, United Kingdom, pp. 11-48.

Pearce, D. and E. Barbier (1991), The Greenhouse Effect: A View from Europe, *The Energy Journal*, Vol. 12, No. 1, pp. 147-60.

Pearce, D. and J. Warford (1993), *World without End: Economics, Environment, and Sustainable Development*, Oxford University Press, Oxford.

Pearson, P.J.G. (1989), Proactive Energy-Environment Policy Strategies: A Role for Input-Output Analysis?, *Environment and Planning A*, Vol. 21, pp. 1329-48.

Peck, S.C. and T.J. Teisberg (1993), CO_2 Emissions Control: Comparing Policy Instruments, *Energy Policy*, Vol. 21, No. 3, pp. 222-30.

Peerlings, J. (1993), *An Applied General Equilibrium Model for Dutch Agribusiness Policy Analysis*, PhD Dissertation, University of Wageningen, Wageningen, The Netherlands.

Pereira, A. and J. Shoven (1988), Survey of Dynamic General Equilibrium Models for Tax Policy Evaluation, *Journal of Policy Modeling*, Vol. 10, No. 3, pp. 401-36.

Pezzey, J. (1992), Analysis of Unilateral CO_2 Control in the European Community and OECD, *The Energy Journal*, Vol. 13, No. 3, pp. 159-71.

Piggott, J., Whalley, J. and R. Wigle (1992), International Linkages and Carbon Reduction Initiatives, in K. Anderson and R. Blackhurst (eds), pp. 115-29.

Piggott, J., Whalley, J. and R. Wigle (1993), Carbon-Reduction Initiatives? *Journal of Policy Modeling*, Vol. 15, Nos. 5&6, pp. 473-90.

Poterba, J. M. (1993), Global Warming Policy: A Public Finance Perspective, *Journal of Economic Perspectives*, Vol. 7, No. 4, pp. 47-63.

Proops, J.L.R., Faber, M. and G. Wagenhals (1993), *Reducing CO_2 Emissions: A Comparative Input-Output Study for Germany and the UK*, Springer-Verlag, Berlin.

Proost, S. and D. van Regemorter (1992), Economic Effects of a Carbon Tax: With a General Equilibrium Illustration for Belgium, *Energy Economics*, Vol. 14, No. 2, pp. 136-49.

Proost, S. and D. van Regemorter (1994), *Testing the Double Dividend Hypothesis for a Carbon Tax in a Small Open Economy*, Centre for Economic Studies, Catholic University of Leuven, Leuven, Belgium.

Pyatt, G. and J. Round (eds) (1985), *Social Accounting Matrices: A Basis for Planning*, The World Bank, Washington, DC.

Qu, Geping (1992a), China's Dual-Thrust Energy Strategy: Economic Development and Environmental Protection, *Energy Policy*, Vol. 20, No. 6, pp. 500-6.

*Qu, Geping (1992b), *Environment and Development in China*, China Environmental Science Press, Beijing.

Ramakrishna, K. (ed., 1994), *Criteria for Joint Implementation under the Framework Convention on Climate Change*, Report of a Workshop held January 9-11, Woods Hole Research Center, Massachusetts.

Robinson, S. (1989), Multisectoral Models, in H. Chenery and T.N. Srinivasan (eds), *Handbook of Development Economics*, Vol. 2, pp. 885-947, North-Holland, Amsterdam.

Robinson, S., Kenneth, H. and M. Kilkenny (1990), *The USDA/ERS Computable General Equilibrium (CGE) Model of the United States*, Staff Report No. AGES 9049, Economic Research Service, U.S. Department of Agriculture, Washington, DC.

Roland-Holst, D., Lee, H. and D. van der Mensbrugghe (1993), Modeling Chinese Growth and Trade from a General Equilibrium Perspective, Prepared for the Development Research Center of the State Council of the P.R. China, Beijing.

Rose, A. (1990), Reducing Conflict in Global Warming Policy: The Potential of Equity as a Unifying Principle, *Energy Policy*, Vol. 18, No. 12, pp. 927-35.

Rose, A. and B. Stevens (1993), The Efficiency and Equity of Marketable Permits for CO_2 Emissions, *Resources and Energy Economics*, Vol. 15, No. 1, pp. 117-46.

Rose, A., Tompkins, M.M., Lim, D., Frias, O. and J. Benavides (1994), *Coal Use in the People's Republic of China, Vol. 2: The Economic Effects of Constraining Coal Utilization*, ANL/DIS/TM-22, Argonne National Laboratory, Argonne, Illinois.

Rosenberg, N.J., Easterling, W.E.III., Crosson, P.R. and J. Darmstadter (eds) (1988), *Greenhouse Warming: Abatement and Adaptation*, Proceedings of a Workshop held in Washington, DC June 14-15, Resources for the Future, Washington, DC.

Scarf, H. (1967), On the Computation of Equilibrium Prices, in W.J. Feliner (ed), *Ten Economic Studies in the Tradition of Irving Fisher*, Wiley, New York.

Scarf, H. (1984), The Computation of Equilibrium Prices, in H. Scarf and J. Shoven (eds), pp. 1-50.

Scarf, H. and T. Hansen (1973), *The Computation of Economic Equilibria*, Yale University Press, New Haven.

Scarf, H. and J. Shoven (eds)(1984), *Applied General Equilibrium Analysis*, Cambridge University Press, Cambridge.

Schelling, T.C. (1992), Some Economics of Global Warming, *American Economic Review*, Vol. 82, No. 1, pp. 1-14.

Shah, A. and B. Larsen (1992), *Carbon Taxes, the Greenhouse Effect, and Developing Countries*, Policy Research Working Papers No. 957, The World Bank, Washington, DC.

Shaw, R.W. (1993), Acid-Rain Negotiations in North America and Europe: A Study in Contrast, in G. Sjöstedt (ed), *International Environmental Negotiation*, SAGE, Newbury Park, California.

*Shen, D.S., Miao, T.J. and X.H. Feng (1992), The Strategic Position of Energy Conservation Planning in China, in CERS (ed, 1992), pp. 27-31.

Shi, Xia (1991), *Modeling the Chinese Economy: In a General Equilibrium Framework*, PhD Dissertation, Stanford University.

*Shi, Xiangjun et al. (1992), A Comparison of Future Nuclear Power and Coal-fired Power in China, *Nuclear Economics*, Vol. 2, pp. 74-83.

Shoven, J.B. and J. Whalley (1984), Applied General-Equilibrium Models of Taxation and International Trade: An Introduction and Survey, *Journal of Economic Literature*, Vol. 22, pp. 1007-51.

Shoven, J.B. and J. Whalley (1992), *Applying General Equilibrium*, Cambridge University Press, Cambridge.

Shunker, A., Salles, J.-M. and C. Rios-Velilla (1992), Innovative Mechanisms for Exploiting International CO_2-Emission Abatement Cost differences, *European Economy*, Special Edition No. 1, pp. 299-338.

Singh, I. (1992), *China: Industrial Policies for an Economy in Transition*, World Bank Discussion Papers No. 143, The World Bank, Washington, DC.

Skea, J. (1995), Environmental Technology, in H. Folmer, L. Gabel and J. Opschoor (eds), *Principles of Environmental and Resource Economics: A Guide for Students and Decision-Makers*, Edward Elgar, Aldershot, England, pp. 389-412.

Smith, C. (1989), *Integrated Multiregion Models for Policy Analysis: An Australian Perspective*, North-Holland, Amsterdam.

Smith, S. (1992), *Distributional Effects of a European Carbon Tax*, Nota Di Lavoro 22.92, Fondazione Eni Enrico Mattei, Milano.

Standaert, S. (1992), The Macro-Sectoral Effects of an EC-Wide Energy Tax, in F. Laroui and J.W. Velthuijsen (eds), pp. 27-63.

*State Economic and Trade Commission (1995), *China Energy Annual Review 1994*, Beijing.

*State Economic and Trade Commission (1996), *China Energy Annual Review 1996*, Beijing.

*State Information Center (1993), Preliminary Forecast for China's Economic Development in the Early 21st Century, *Economic Forecast and Analysis*, No. 25, Beijing.

*State Planning Commission (1992), *A Study on Greenhouse Gases Emissions: Alternative Energy Supply Options to Substitute for Carbon-Intensive Fuels*, Beijing.

State Planning Commission (1993), *Alternative Energy Supply Options to Substitute for Carbon-Intensive Fuels: A Review of Scenario Analysis*, Beijing.

*State Statistical Bureau (1986), *Input-Output Table of China 1981*, Economic Forecasting Center of the State Planning Commission and Department of Balances of National Economy of the State Statistical Bureau, State Statistical Publishing House, Beijing.

*State Statistical Bureau (1987), *Energy Statistical Yearbook of China 1986*, State Statistical Publishing House, Beijing.

*State Statistical Bureau (1989), *Statistical Yearbook of China 1989*, State Statistical Publishing House, Beijing.

*State Statistical Bureau (1990a), *Almanac of China's Economy 1990*, State Statistical Publishing House, Beijing.

*State Statistical Bureau (1990b), *Industrial Economic Statistical Yearbook of China 1990*, State Statistical Publishing House, Beijing.

*State Statistical Bureau (1990c), *Energy Statistical Yearbook of China 1989*, State Statistical Publishing House, Beijing.

*State Statistical Bureau (1991), *Input-Output Table of China 1987*, Department of Balances of National Economy of the State Statistical Bureau and Office of the National Input-Output Survey, State Statistical Publishing House, Beijing.

*State Statistical Bureau (1992a), *Statistical Yearbook of China 1992*, State Statistical Publishing House, Beijing.

*State Statistical Bureau (1992b), *Energy Statistical Yearbook of China 1991*, State Statistical Publishing House, Beijing.

Stephan, G. (1992), Environmental Regulations and Innovation: A CGE Approach for Analysing Short-Run and Long-Run Effects, in T. Sterner (ed.), pp. 299-321.

Stephan, G., van Nieuwkoop, R. and T. Wiedmer (1992), Social Incidence and Economic Costs of Carbon Limits: A Computable General Equilibrium Analysis for Switzerland, *Environmental and Resource Economics*, Vol. 2, pp. 569-591.

Sterner, T. (ed)(1992), *International Energy Economics*, Chapman & Hall, London.

Stocks, K. and A. Musgrove (1984), MENSA - A Regionalized Version of MARKAL, The IEA Linear Programming Model for Energy System Analysis, *Energy Systems and Policy*, Vol. 8, No. 4, pp. 313-48.

Sun, M. (1990), Emissions Trading Goes Global, *Science*, Vol. 247, pp. 520-21.

Symons, E., Proops, J. and P. Gay (1994), Carbon Taxes, Consumer Demand and Carbon Dioxide Emissions: A Simulation Analysis for the UK, *Fiscal Studies*, Vol. 15, No. 2, pp. 19-43.

TERI and INET (1990), *Energy Development in China and India: A Comparative Study of Energy Supply, Energy Consumption and Energy Policies*, Tata Energy Research Institute (TERI), New Delhi; Institute of Nuclear Energy Technology (INET), Beijing.

Tietenberg, T.H. (1990), Economic Instruments for Environmental Regulation, *Oxford Review of Economic Policy*, Vol. 6, No. 1, pp. 17-33.

Tietenberg, T. (1992), *Environmental and Natural Resource Economics*, Harper Collins, New York, 3rd Edition.

Tietenberg, T. and D.G. Victor (1994), Possible Administrative Structure and Procedures (for Implementing a Tradeable Entitlement Approach to Controlling Global Warming), in United Nations Conference on Trade and Development, *Combating Global Warming: Possible Rules, Regulations and Administrative Arrangements for a Global Market in CO_2 Emission Entitlements*, Geneva, pp. 1-60.

Treadwell, R., Thorpe, S. and B.S. Fisher (1994), Should We Have Second Thoughts about 'No-Regrets'?, in OECD (1994), pp. 255-72.

Turner, R. Kerry, Pearce, D. and I. Bateman (1994), *Environmental Economics: An Elementary Introduction*, Harvester Wheatsheaf, New York.

United Nations (1993), *World Economic Survey: Current Trends and Policies in the World Economy*, New York.

UNEP (1988), The Changing Atmosphere: Implications for Global Security, *Conference Statement*, Toronto, Canada, June 27-30.

Varian, H.R. (1992), *Microeconomic Analysis*, Norton, New York, 3rd Edition.

Varian, H.R. (1993), *Intermediate Microeconomics: A Modern Approach*, Norton, New York, 3rd Edition.

Virdis, M.R. and M. Rieber (1991), The Cost of Switching Electricity Generation from Coal to Nuclear Fuel, *The Energy Journal*, Vol. 12, No. 2, pp. 109-34.

Walker, I.O. and F. Birol (1992), Analysing the Cost of an OECD Environmental Tax to the Developing Countries, *Energy Policy*, Vol. 20, No. 6, pp. 568-74.

*Wang, Jingxin (1992), The Importance of Natural Gas in China's Energy System and Related Issues about Its Development Strategy, in CERS (ed, 1992), pp. 108-20.

Ward, W.A. and J.B. London *et al.* (1994), *Energy Efficiency in China: Case Studies and Economic Analysis*, Subreport No. 4 for China GHG Study, World Bank, Washington, DC.

Watt, E., Sathaye, J. and O. de Buen *et al.* (1995), *The Institutional Needs of Joint Implementation Projects*, LBL-36453, Energy and Environment Division, Lawrence Berkeley Laboratory, Berkeley, California.

*Wei, Y.J., Du, L.F. and X.Q. Yang (eds)(1989), *China's Current Tax System*, People's University Press, Beijing.

Weitzman, M.L. (1974), Prices vs. Quantities, *Review of Economic Studies*, Vol. 41, pp. 477-91.

Welsch, H. (1993), Economic Approaches to International Agreement for Carbon Dioxide Containment, *International Journal of Global Energy Issues*, Vol. 5, Nos. 2/3/4, pp. 114-22.

*Wen, Hongjun (1992), Exploration for Cost of Constructing Nuclear Power Plants in China, *Nuclear Economics*, Vol. 2, pp. 58-62.

Weyant, J.P. (1993), Costs of Reducing Global Carbon Emissions, *Journal of Economic Perspectives*, Vol. 7, No. 4, pp. 27-46.

Whalley, J. (1991), The Interface between Environmental and Trade Policies, *The Economic Journal*, Vol. 101, pp. 180-89.

Whalley, J. and R. Wigle (1991a), Cutting CO_2 Emissions: The Effects of Alternative Policy Approaches, *The Energy Journal*, Vol. 12, No. 1, pp. 109-24.

Whalley, J. and R. Wigle (1991b), The International Incidence of Carbon Taxes, in R. Dornbusch and J. Poterba (eds), pp. 71-97.

Williams, R.H. (1990), Low-Cost Strategies for Coping with CO_2 Emission Limits (A Critique of 'CO_2 Emission Limits: An Economic Cost Analysis for the USA' by Alan Manne and Richard Richels), *The Energy Journal*, Vol. 11, No. 3, pp. 35-59.

Williams, R.H. (1992), The Potential for Reducing CO_2 Emissions with Modern Energy Technology: An Illustrative Scenario for the Power Sector in China, *Science and Global Security*, Vol. 3, pp. 1-42.

Wilson, D. and J. Swisher (1993), Exploring the Gap: Top-down versus Bottom-up Analyses of the Cost of Mitigating Global Warming, *Energy Policy*, Vol. 21, No. 3, pp. 249-63.

World Bank (1985a), *China: The Energy Sector, Annex 3 to China: Long-Term Development Issues and Options*, A World Bank Country Study, Washington, DC.

World Bank (1985b), *China: Economic Model and Projections, Annex 4 to China: Long-Term Development Issues and Options*, A World Bank Country Study, Washington, DC.

World Bank (1985c), *China: Economic Structure in International Perspective, Annex 5 to China: Long-Term Development Issues and Options*, A World Bank Country Study, Washington, DC.

World Bank (1988a), *World Development Report 1988*, Oxford University Press, Oxford.

World Bank (1988b), *China: External Trade and Market*, Washington, DC.

World Bank (1990a), *China: Between Plan and Market*, Washington, DC.

World Bank (1990b), *China: Financial Sector and Institutional Development*, Washington, DC.

World Bank (1990c), *China: Revenue Mobilization and Tax Policy*, Washington, DC.

World Bank (1990d), *China: Macroeconomic Stability and Industrial Growth under Decentralized Socialism*, Washington, DC.

World Bank (1992a), *World Development Report 1992: Development and the Environment*, Oxford University Press, Oxford.

World Bank (1992b), *World Tables 1992*, The Johns Hopkins University Press, Maryland.

World Bank (1994), *China Power Sector Reform: Toward Competition and Improved Performance*, China and Mongolia Department Report No. 12929-CHA, Washington, DC.

World Resources Institute (1990), *World Resources 1990-91*, Oxford University Press, New York.

World Resources Institute (1992), *World Resources 1992-93*, Oxford University Press, New York.

Wu, Jinglian and Renwei Zhao (1987), The Dual Pricing System in China's Industry, *Journal of Comparative Economics*, Vol. 11, No. 3, pp. 309-18.

Wu, Zongxin and Zhihong Wei (1992), 'China', in R.K. Pachauri and P. Bhandari (eds), *Global Warming: Collaborative Study on Strategies to Limit*

CO₂ Emissions in Asia and Brazil, Chapter 4, Tata McGraw-Hill, New Delhi, pp. 45-63.

Wu, Shiyu (1995), Coal Preparation in China - Current Situation and Outlook, *Aufbereitungs-Technik*, Vol. 36, No. 8, pp. 353-6.

Xia, G. and Z.H. Wei (1994), Climate Change and the Technical and Institutional Adaptations: China's Perspective, in N. Nakicenovic, W.D. Nordhaus, R. Richels and F.L. Toth (eds), *Integrative Assessment of Mitigation, Impacts, and Adaptation to Climate Change*, IIASA CP-94-9, Austria, pp. 411-25.

*Yan, Changle (ed)(1994), *China's Energy Development Report*, Economic Management Press, Beijing.

Yang, Hongnian (1991), Coal Transportation System and Assessment of Environmental and Economic Implication in China, in *Environmentally Sound Coal Technologies: Policy Issues and Options*, Report of the International Conference on Coal and Environment, Beijing, December 2-6, pp. 148-56.

*Yang, Xiaosheng (1992), The Development of Renewable Energy, *Energy of China*, No. 3, pp. 18-20.

*Yang, Zhirong (1992), Suggestions on the Strategies and Policies of Hydropower Exploitation in China, *Bulletin of Energy Policy Research*, No. 8, pp. 11-7.

*Yao, Yufang et al. (1993), *Outlook for China Economic and Social Development*, A Draft Report, Institute of Quantitative and Technical Economics, Chinese Academy of Social Sciences, Beijing.

*Yao, Yufang et al. (1994), *Outlook for China Economic and Social Development*, Two Volumes, Institute of Quantitative and Technical Economics, Chinese Academy of Social Sciences, Beijing.

*Ye, Qing (1989), Summary of the 5th Official Meeting of the Energy Conservation of the State Council, *Energy of China*, No. 5, pp. 4-8.

Zhai, Yongping (1993), Carbon Dioxide Emissions from Fossil Fuels in China: Historical Analysis (1980-1990) and Prospects (2000-2015), *International Journal of Global Energy Issues*, Vol. 5, Nos. 2/3/4, pp. 177-85.

Zhang, ZhongXiang (1991), *Evolution of Future Energy Demands and CO₂ Emissions up to the Year 2030 in China*, ECN-I-91-038, Netherlands Energy Research Foundation (ECN), Petten, The Netherlands.

Zhang, ZhongXiang (1992) *The MARKAL Model*, Department of General Economics, University of Wageningen, Wageningen, The Netherlands.

Zhang, ZhongXiang (1994a), Setting Targets and the Choice of Policy Instruments for Limiting CO₂ Emissions, *Energy and Environment*, Vol. 5, No. 4, pp. 327-41.

Zhang, ZhongXiang (1994b), Analysis of the Chinese Energy System: Implications for Future CO₂ Emissions, *International Journal of Environment and Pollution*, Vol. 4, Nos. 3/4, pp. 181-98.

Zhang, ZhongXiang (1995a), Economic Approaches to Cost Estimates for Limiting Emissions of Carbon Dioxide, *International Journal of Environment and Pollution*, Vol. 5, Nos. 2/3, pp. 194-203.

Zhang, ZhongXiang (1995b), Energy Conservation in China: An International Perspective, *Energy Policy*, Vol. 23, No. 2, pp. 159-66.

*Zhang, ZhongXiang (1995c), Alternative Economic Approaches to Estimating the Costs of Carbon Dioxide Mitigation, *Bulletin of Energy Policy Research*, Vol. 16, No. 6, pp. 10-5.

Zhang, ZhongXiang (1996a), Energy, Carbon Dioxide Emissions, Carbon Taxes and the Chinese Economy, *Intereconomics (Review of International Trade and Development)*, Vol. 31, No. 4, pp. 197-208.

Zhang, ZhongXiang (1996b), Some Economic Aspects of Climate Change, *International Journal of Environment and Pollution*, Vol. 6, Nos. 2/3, pp. 185-95.

Zhang, ZhongXiang (1997a), Cost-Effective Analysis of Carbon Abatement Options in China's Electricity Sector, An Invited Paper for *Energy Sources*, Special Issue on Energy, Environment and Sustainable Development, Vol. 19.

Zhang, ZhongXiang (1997b), Macroeconomic Effects of CO_2 Emission Limits: A Computable General Equilibrium Analysis for China, *Journal of Policy Modeling*, Vol. 19, No. 5.

Zhang, ZhongXiang (1997c), Macroeconomic and Sectoral Effects of Carbon Taxes: A General Equilibrium Assessment for China, *Economic Systems Research*, (forthcoming).

Zhang, ZhongXiang (1997d), Operationalization and Priority of Joint Implementation Projects, *Intereconomics (Review of International Trade and Development)*, Vol. 32; A Shortened Version of the Report *Joint Implementation as a Cost-Effective Climate Policy Measure: A Chinese Perspective* Prepared for the Dutch Ministry of Housing, Spatial Planning and the Environment under Contract 95140042.

*Zhao, Zhijun (1989), A Comparative Analysis of Hydroelectric Power, Coal-fired Power and Nuclear Power, *Bulletin of Energy Policy Research*, Vol. 10, No. 10, pp. 7-14.

*Zhao, Dianwu and Baozhong Wu (1992), Control Technology and Policy of Energy Related Environmental Pollution in China, in Qingyi Wang (ed), *Energy and Environment*, Ministry of Energy, Beijing, pp. 22-7.

Zhou, D.D. and J.F. Li (1995), Case Study of China, in M. Mabel, E. Watt and J. Sathaye (eds), *Perspectives on the Institutional Needs of Joint Implementation Projects for China, Egypt, India, Mexico, and Thailand*, LBL-37031, Energy and Environment Division, Lawrence Berkeley Laboratory, Berkeley, California, pp. 43-53.

Zhou, F.Q., Qu, S.Y. and T. Rong (1989), *The Sectoral Energy Demand Analysis in China: Utilization of the MEDEE-S Model*, Energy Publishing House, Beijing.

Zhou, Fengqi (1992), Comments on 'The Potential for Reducing CO_2 Emissions in China with Modern Technology', *Science and Global Security*, Vol. 3, pp. 43-7.

*Zhu, Chengzhang (1992), The Present Situation of China's Energy Policy and Its Reform, in CERS (ed, 1992), pp. 133-40.

*Zhu, Shiwei et al. (1991), A Study on Evaluating Wind Electric Power, *Quantitative and Technical Economics*, Vol. 8, No. 11, pp. 37-44.

Index